AF808694

BIOLOGY TOPIC TEXT A: SCIENCE, LIFE AND BASIC CHEMISTRY FOR BIOLOGY STUDENTS

HAROLD M. ICKES II, M.D.

Executive Editor: Rhett C. Smith, Ph.D.

Ideal Quill Publishing Group LLC

idealquill.com

Printed in the United States of America

10 9 8 7 6 5 4 3 2 1

ISBN 978-0615747439

Biology Topic Text A: Science, Life and Basic Chemistry for Biology Students

by Harold M. Ickes II, M.D.

Table of Contents

Unit 1

The Science of Biology

In this section we cover:
1) The definition of biology
2) The basic concepts of science

What is Biology?

Biology is the science that studies life. Unfortunately, that definition uses the word 'life,' which can be very problematic. What constitutes 'life' is not so easily defined, because the processes of life span a huge range of forms. It comes in practically all shapes and sizes. At times, it's hardly recognizable from nonliving things, and nonliving things can at times appear to be alive. At other times, we can recognize it intuitively, from the mailman at your door, to the cat walking across your yard, to the strange thing currently growing in my kitchen sink (put there by my lazy roommate, no doubt). If we are to study and categorize life, clearly we must first specify the difference between 'living' and 'nonliving.' To do this, we have to be very systematic and careful. We need to be scientific. To do that, it is important to understand the meaning of 'science.' While it may seem silly to toil over the meaning of science, it is useful to have a clear definition in mind. All too often, the media is full of debates about what science really is and what it isn't. So, let us talk about this for a little bit.

What is science and why does it exist?

Science is a field of human endeavor that seeks to attain knowledge through a systematic approach of gathering evidence. The knowledge we acquire is then used to predict further outcomes. Why do we do this? Often, we seek evidence in nature because we are just plain curious. Just as often, we seek out answers because we want to predict something that will make us rich or famous. For example, what natural laws could help me make a better cell phone? What biological components are out there to allow me to help prevent acne? How could I make a floating car? How can I use science to rid me of my neighbor's terrible music? All of these things benefit mankind and help to make the scientist rich. There's a catch: in order to gain benefit or wealth from your research, your theories have to be proven to work. Theories are tested and tested and tested and tested for the slightest weakness. When that process is done, it's tested again. If something about a theory appears to falter, millions of people stand ready to point out the weakness and propose a better theory, especially if the new theory leads to better cell phones or acne preventions or floating cars. That is the business of science, and no accepted scientific theory exists by virtue of it being cool or trendy. Our theories are in place because we can't find holes in them, but rest assured we test them all the time. As long as people are curious, in need of improving the human condition, or simply want to get rich, our theories will be rigorously tested.

Three Assumptions about Reality

Ok, how do we do science? We do science using the **scientific method**. This method is quite simple and surprisingly easy to do, but we first have to understand three assumptions about reality. They may seem obvious to you, but it's important that these assumptions be made. The assumptions are: 1) **Natural Causality**, 2) **Uniformity of Space-Time**, and 3) the assumption of **Common Sensation**. These three principles, together with the scientific method, allow us to slowly unravel the causes of Nature.

First, let's talk about **Natural Causality**. What does it mean? Well, in history, there were two ways of trying to explain Nature. The first was to suppose a divine origin. The thing in my kitchen sink exists, one might say, because a divine being such as God put it there. The second way to explain nature was through Natural Causality, **a principle that assumes that all things occur as a consequence of some other natural event that can be understood** (although not necessarily *easily* understood). Natural Causality is basically a way of saying: "let's see how much we can understand about our reality without assuming anything supernatural." It is not a statement about the existence of God nor an attempt to disprove Him (Her?). **Disproving God is *not* the goal of science**. We are simply trying our best to understand Nature before shrugging and saying that God did it.

Therein lies a subtlety, however, for when do we ever know if something cannot be explained by natural events? Perhaps we haven't looked deep enough, after all. The answer may be around the corner, so in effect, Natural Causality stops us from *ever* making any statement about a supernatural origin of any phenomena. We assume an answer exists and keep digging.

The second assumption is the **Uniformity of Space-Time**. What is that? Is it some mad and convoluted monstrosity from the mind of Einstein? No, not at all. It just means that **we assume the laws of physics work the same everywhere in the Universe and have worked the same throughout all time**. Imagine how incredibly impossible it would be to try to figure anything out if the laws of physics were completely different depending on the era or the place. Now THAT would be mad and convoluted. For example, suppose your car stops working and you're trying to figure out what is wrong with it. Now suppose the laws of physics completely change every 3rd second. Nothing would ever make sense and you'd never get your car working. (I doubt cars would ever have been invented in that case, actually). What if the laws of physics change when you go from one state or country to another? Your car would stop altogether because the physics that runs it would change upon crossing a border.

The third principle is the assumption of "**Common Sensation**." No, I didn't say common sense, which is anything but common. "Common Sensation" **means that we all observe the world the same**. We see, hear, feel, smell and taste things in the same fashion. In other words, if we are both looking at a square, we will both say it is a square. It won't mystically look like a circle to you for some reason. If we are both looking at the color blue, we will both comment it is blue. What is hot to me is hot to you, and so forth. This principle speaks about things we can measure physically; we must be able to agree that 5 meters is 5 meters or the very nature of measurement is sketchy, and then no one can get rich doing science.

The assumption of common sensation is not referring to how we feel about those things. So, for example, you and I could watch the same movie, but I might like the movie while you might not. We may disagree about the quality of the movie, but we can readily agree that we saw the same movie, with the same actors, scenes, action, and so forth. How we feel about things, such as movies, politics, religion, etc., has no formal place in science. We all may have opinions about scientific theories, but unless we can *show* that these theories are wrong, our opinions are just opinions. Your opinion cannot be made into a theory, unless it is well formulated with mountains of rock-solid evidence behind it.

Note that the assumption of common sensation requires a healthy and normally functioning mind. There are a number of unfortunate illnesses that cause people to be unable to accurately sense and decipher the physical world. As a result, they often suffer from being unable to understand what is real and what is not.

> **Important Founding Principles of Science**
>
> 1) Natural Causality – phenomena have natural explanations that can be understood.
> 2) Uniformity of Space-Time – the laws of physics work the same throughout the Universe and throughout all time.
> 3) Common Sensation – we all observe the physical world in the same ways.

The Scientific Method

The scientific method is a systematic approach to investigating reality. It begins with the **observation**. From that observation, a **hypothesis** is made. A hypothesis is an educated guess, (unless you are my housemate, in which case it would just be a guess). The hypothesis is only useful if it can be tested. For example, suppose I declare that I have an untouchable, unseeable, undetectable Doobadooboo monster in my closet. If it is impossible to detect it in any way, then it is silly to even argue about whether or not it is there. Science deals with the testable, and if your theories cannot be tested, it is not science.

Once we have a testable hypothesis, the next thing to happen is to **experiment** to verify or disprove the hypothesis. When we do this, we do physical measurements, once under one condition and then again under a different condition. When changing the conditions, it is best to change only one thing at a time. For example, suppose you want to understand the effects of pressure on the boiling point of water. In one situation, you boil the water under conditions of standard room pressure. In the second situation, you do *exactly the same thing*, changing *only* the pressure in the room. Assuming nothing else changed, the change in pressure can then be blamed for whatever change happens to the boiling point. Suppose, on the other hand, you *also* changed the pot as well. Could you be certain that the change in boiling point was due to the change in pressure and not the change in the pot? Sadly, you couldn't. So, to avoid that, we change *only one thing at a time*. This is an important point to understand. Before your idea will become an accepted theory, it will be attacked from all directions by people who will argue that more than one thing changed. You must be able to demonstrate that only one thing changed between the two conditions. Sometimes it is trickier than it sounds, especially in biology and related fields like medicine.

Lastly, after the experiment, we make a **conclusion** by accepting, altering, or rejecting our hypothesis. If the hypothesis fails experimentation, we come up with a new hypothesis and start over.

Summary

So, to summarize this nifty little chapter, we learned about the science of biology, the nature of science, and the laziness of my housemate. In future sections, we'll build upon these ideas.

- What is biology?
    - It is the science that studies life.
- What are the three assumptions before studying any science?
    - Natural Causality – phenomena have natural explanations that can be understood.
    - Uniformity of Space-Time – the laws of physics work the same throughout the Universe and throughout all time.
    - Common Sensation – we all observe the physical world in the same ways.
- What is the scientific method?
    - 1) Observation – we observe something that we don't understand.
    - 2) Hypothesis – we make an educated and testable guess about our observation.
    - 3) Experiment – we test the hypothesis by changing one parameter of the experiment at a time.
    - 4) Conclusion – we accept, alter, or reject the original hypothesis, returning to (2) if need be.

Unit 2

What is Life?

In this section we cover:
1) The definition of life, focusing on an overview of its layered structure, from particles on up to the biosphere.
2) Taxonomy

All right, enough grumbling about what science is. Now that we have made those introductory statements, let us clarify what we mean by the difference between living and nonliving. In other words, let's define what life is. All living things have eight traits in common: **complex structure**, **response to stimuli**, **homeostasis**, **metabolism**, **growth**, **reproduction**, **inheritable genetics** and **evolution**. There were a couple unfriendly words in there, so let's discuss these a little more in depth. Once we've done that, we'll briefly discuss how life is categorized.

Complex Structure

Living organisms are incredibly complex things, made up of very complex structure. Biology divides these structures into various levels of organization, and in this Topics Text we will focus on the smaller aspects. Still, it is important to have an idea of the greater picture, so here is a glance at how life is structured from smallest to largest pieces.

The smallest entities in the known Universe are **particles**. Particles are (probably) indivisible – meaning that we cannot break them down into smaller things, as far as we know. Particles come together to make **atoms**. Atoms, in turn, clump together in various fashions to make **molecules**. Molecules that have carbon in them are called organic molecules or organic compounds[1]. Organic compounds

---

[1] It is very unfortunate that society has chosen the word "organic" to denote a food that was naturally grown. All food is, by its very nature, organic since it has carbon in it. If it were not made with carbon it would be inorganic and probably poisonous to you.->

are important in biology, because they are the predominant molecules in living systems. We will be discussing them in future units until we are blue in the face. The next level of biological organization is the **organelle**. These are collections of organic molecules put together in a very specific way. Organelles have many purposes in life, but we will put off their functions until a later book. For the moment, think of them as little chemical factories.

One more level up is the **cell**. The cell is the basic unit of life. Cells contain organelles, water with organic molecules floating in it, and a whole lot of other important things needed for life's processes. Think of cells as little water balloons with tiny things floating inside. The boundary of the cell is comprised of the cell membrane, which is like its skin, in a sense. Cells come in all shapes and sizes. Some are amorphous (like blobs, oozes, or my roommate). Some are long and slender, like nerve cells. Some are like cubes. The shape of the cell depends on what its purpose is, as will become evident in future studies. Most life forms have only one cell, like bacteria for example. Bacteria are just simple cells living on their own in water. They can be rod-shaped, long and branched, or ball-shaped. Though they are often living on their own, many can clump together into little colonies.

For those organisms that are made up of two or more kinds of cells, we must go further in our categorization of the levels of organization. A collection of different types of cells (that cannot survive on their own) makes up a **tissue**. There are many types of tissues, such as nervous tissue, muscle tissue and connective tissue, to name a few.

Next on the list of organization levels is the organ. **Organs** are made up of two or more types of tissues. The heart, the brain, the pancreas and the liver are examples.

Organ systems come next. The brain connected to the spinal cord (another organ) make up the nervous system; the heart connected to all the blood vessels make up the circulatory system, etc.

A collection of organ systems makes up an individual, or the **multicellular organism**. This is the level we readily recognize in

---

Actually – come to think of it – it is also unfortunate that we say something is "naturally" grown. After all, I can't remember the last time I saw something that was supernaturally grown.

everyday life. When we greet each other, we see each other as complete individuals. We never see a brain moseying in to the post office by itself[2], or spot a collection of connective tissue out for a day's fishing. On the other hand, not all organisms necessarily have all organs. Trees do not have hearts, squids do not have knee caps, and my roommate has managed very well without a brain.

One might suppose that our discussion of the structure of life would end here, but it does not. Biology concerns much more than this. A collection of individuals living within the same area and capable of interbreeding is called a **population**. These individuals tend to share very similar traits, since they interbreed with each other. The humans living in a city on an isolated island (say before boats and planes) would be a population, because they can interbreed with each other and tend to be very similar. They might, for example, all have blond hair with many of them having green eyes. Who knows?

A collection of populations, all capable of interbreeding with each other, is known as a **species**. Two populations could be on the opposite side of the Earth from each other, but if they are capable of breeding with each other, they are part of the same species. For example, the indigenous peoples of South America are one population and the people of Japan are another, but since they are capable of breeding and producing viable reproductive offspring, the two populations are considered part of the same species. If a *healthy male* and a *healthy female* are *not* capable of breeding together, they are *not* part of the same species. If those two organisms can breed together and the child can (eventually) reproduce as well, then the two parents are part of the same species. For a quick and rough definition: Interbreeding potential = same species.

Next on the list is the **community**. A community is defined as two or more species, such as a species of frog, a species of fish and a species of mosquito, living in the same area.

The **ecosystem** is one level higher. It is the community with the nonliving elements considered as well. So, the frogs, the fish, the mosquitoes, the crazy guy at the end of the street with the antler hat... along with the rocks, the streams, etc. make up the ecosystem in which those species live.

---

[2] ...sauntering perhaps, but never moseying.

Last on the list (phew!), is the **biosphere**. It basically means the part of the Earth in which any form of life can be found. There is only one biosphere until we find another planet with life on it (or put life on it), or until we mess up everything and kill the one we're on. Oh glorious future!

Quick list of biological organization:

Particle – The tiniest known pieces of matter in the Universe.
Atom – A collection of particles put together in a very specific way.
Molecule – A collection of atoms put together in a very specific way.
Organelle – A collection of molecules that have specific purposes in living reactions.
Cell – The basic unit of life, containing many organelles.
Tissue – A collection of two or more types of cells, (which could not survive without each other).
Organ – A collection of two or more types of tissues.
Organ System – A collection of two or more organs working in concert.
Multicellular Organism – An individual made up of multiple cells, often arranged as a collection of organ systems. (The generic term "organism" is often used instead. The term organism can, however, refer to any kind of critters, multicellular or not.)
Population – A collection of interbreeding organisms.
Species – A collection of populations, all capable of interbreeding if the opportunity arises.
Community – Collection of different species living in a geographic area.
Ecosystem – The community plus the nonliving elements of the geographic area.
Biosphere – The surface of the Earth.

Responding to Stimuli

Now we know that living things are complex and made up of a tremendous number of carbon-containing molecules known as organic compounds. The next quality that life has is the ability to respond to environmental circumstances that could alter the continued survival of the organism or its offspring. For example, if a cat gets hit with a stick, the cat will jump away. Why? The impact of the stick upon the cat causes pain to the creature, which is a built-in response that tells the cat "this is causing damage that could threaten your life and/or ability to reproduce." So, the cat jumps away to avoid being hit again.

Now suppose I hit a small rock with a stick. It appears to "jump" away as well. Does that mean that the rock is fleeing to save its life? No, of course not. The rock flew away because I imparted momentum upon it. It's a simple physics phenomenon. After all, suppose I hit a *big* rock with a stick. The rock will just sit there with absolutely no response, even though continued strikes could (in theory) "threaten" the continued existence of the rock.

That's all well and good, but what if the cat is drugged so that it is incapable of responding to being hit by a stick? Can we then say that the cat is no longer alive because it cannot respond to stimuli? No. We cannot make such a conclusion, because we know the cat would respond under standard conditions. Just because the cat is *stoned* doesn't mean it's the same as the *rock*. . .

Yeah, that's right. I actually went all that way just to set up a terrible joke. As it turns out, sticks and stones break monotony.

Incidentally, I do not, under any circumstance, advocate that you hit cats. Animal cruelty is never advisable, nor acceptable. You are more than welcome to have a go at the thing in my kitchen sink, though. I'm quite certain it will respond to the stimuli, however, and you probably won't like the response. You've been warned.

Homeostasis

Homeostasis is a very important quality of life, and we hinted at it with our example of the cat. The word is derived from Greek and means roughly "stay the same." All living organisms maintain their internal structure to the best of their ability for as long as they can. If they are low on fluids, they get thirsty and seek out water. If they are low on a nutrient, they tend to seek out specific types of food that contain that nutrient. Homeostasis is not just a conscious drive, however, for it happens at the level of the cell as well. If a certain protein (which we will discuss later) is running low within the cell, the cell will gather up the necessary materials and make more of the protein. This quality of maintaining balance is vital to the success of the organism. Without it, life would be like the cat that just sits there – doomed.

Metabolism

Metabolism is the acquisition of material and energy from the environment. These things are manipulated at a chemical level inside the organism and converted into energy and material forms that help it to sustain homeostasis and the other qualities that are required for survival. Think of metabolism as internally controlled chemistry that releases energy and converts scavenged chemicals into more useful forms.

Growth & Reproduction

All living creatures grow. They start small and grow to a bigger size before reproducing. Reproduction takes many forms. Bacteria multiply in number by dividing in half, a means known as **asexual reproduction**. They basically grow until they are twice their original size, then split in half in an orderly fashion to start the cycle anew. Some organisms, such as water hydra, reproduce by **budding** new individuals directly from their sides. In contrast, sexual reproduction occurs when the **gametes** (small genetic packets, if you will) from a male and female combine. Depending on the species, this can occur inside or outside the organism.

Inheritable Genetics

When living organisms reproduce, genetic material is passed from the parent to the child. That genetic material is the DNA[3], a complex molecule that we will discuss in detail later. This molecule stores biological information in it, acting like a blueprint for the new offspring. It dictates such things as the color of your hair (assuming you haven't dyed it), the color of your eyes (minus contacts), your height (to a large extent) and facial features (i.e. family resemblance), to name a few. What we learn (i.e. the skills and most behaviors) is *not* stored in the DNA and is therefore *not* inherited by our children. You are free to learn whatever subject you like, but that particular subject will not be passed to your children. There seems to be plenty of anecdotal evidence that we inherit the *tendency* to learn some subjects quicker than others. In other words, suppose music comes "natural" to your family. One might predict that your daughter would tend to learn piano quicker than the neighbor's rotten kid. It's possible but hard to prove. The problem is that if your daughter is surrounded by music from birth, she'll have picked up much more of it

---

[3] Some viruses use a related molecule called RNA as its genetic material. Whether or not viruses are considered living, however, is hotly debated.

than children of non-musical families, genetic tendency or no. In addition, your experience in music might make you a superior music teacher to her, versus a family with no contacts with musicians. The point is that it's hard to prove that someone is necessarily natural at something, since all the factors may not be evident. Remember, the strongest arguments about cause are those in which only *one* thing is different between two examples. If multiple things in the environment are different, it's hard to pinpoint which is to blame for your family's musical nature.

Either way, there are plenty of things that are definitely NOT inherited, including (and not limited to) scars and most behaviors. These "markings" are not encoded in the DNA. They are acquired during life, and once the organism carrying them dies the markings go with it. If you acquire some psychological injury due to a traumatic event, your children will not be born with the psychological injury, because the details of the psychological injury are not recorded in the DNA.

Basically, the rule of thumb is that *nothing* is inherited unless it is recorded in the DNA, and practically nothing is deliberately written into your DNA as a consequence of your life's experience. Learn to play piano all you want, but your kids won't inherit a scratch of it.

Evolution

Evolution is a complex topic that causes a lot of fuss. Unfortunately, the general populace does not understand the concept very well, because it is poorly taught. The best understanding of the theory of evolution (and what it tells us) comes after we understand a great deal about basic biology, so sadly we will have to wait until a much later date. For the moment, however, let's have a peek at evolution, just for a preview. . .

What is evolution, and why is it so important to understand? It's important because modern biology makes *little sense* without it, and *every bit of sense* with it. The basic idea is simple: inherited information from the DNA contributes to the odds of a creature's survival. Individuals with poor genetics (genes that give them unfavorable physical and basic mental characteristics) have a low chance of survival. Over generations, they fail to reproduce as much as individuals with good genetics and are eventually removed from the species. Individuals with superior genetics have increased odds of survival. They tend to reproduce more successfully, and over the course of generations they increase in percentage.

DNA regularly undergoes **mutations**, which changes the genetic information encoded within an individual. When that individual passes this mutated DNA to an offspring, that offspring may have a new trait that was previously unseen in the species, altering its odds of survival. If this trait is advantageous, the offspring will tend to leave more offspring of its own, and over time the trait will become common. Through the introduction of mutations, new traits appear, and the species slowly changes from one generation to the next. *Individuals do not evolve* (in the scientific meaning of the word). It is a process that takes place in (usually) tiny steps from one generation to the next, sometimes requiring *millions* of years to be obvious. Monkeys did not suddenly climb out of a tree, put on a suit, and decide to walk upright. Well, maybe my housemate did, but he's the only exception.

Anyway, let's discuss a quick example. Suppose a species of bird relies on its ability to break open seeds in order to eat the nutrients inside. Birds born with tiny beaks will have trouble with this task and will have a disadvantage in harvesting nutrients and energy from the seeds. Birds born with larger beaks will be more successful. Each time a mutation appears that makes the beaks a little larger, the birds with these larger beaks tend to do much better than those without. So, they tend to be more likely to survive. Better survival tends to increase the opportunities for mating. Therefore, they tend to mate more. The trait slowly spreads throughout the species. After thousands of years, (or hundreds of thousands), the birds slowly evolve bigger and bigger beaks. The genetics for small beaks are lost because those birds couldn't get enough of the seeds opened.

All the above qualities are required before we can say that something is living. With that covered, we can now move on to the task at hand: basic biological chemistry. Throughout the rest of this book, we will focus on the essential chemistry that makes life go.

Summary

In the unit, we studied nature of life:

- The eight traits all living things have in common are:
    - Complex structure – built up from particles to biosphere, as follows:
        - Particles, atoms, molecules, organelles, cells (unicellular organisms), tissues, organ, organ systems, multicellular organism, population, species, community, ecosystem, biosphere
    - Response to stimuli – altering behavior or state in response to a change in the environment.
    - Homeostasis – the ability to remain relatively unchanged over long period of time as a result of internal chemical maintenance.
    - Metabolism - the ability to chemically alter gathered resources into chemicals not found in the environment or for energy release.
    - Growth - increase in mass and usually volume in preparation for reproduction.
    - Reproduction – the creation of an offspring, either asexually (fission or budding) or sexually (male and female gamete fusing).
    - Inheritable genetics – traits recorded in the genes that get passed on one generation to the next. Scars, injuries (physical or mental), and most behaviors do not get inherited.
    - Evolution – the slow change from one generation to the next due to new mutations and changing environmental pressures. Usually requires thousands of years to notice.

Unit 3

Nonliving Building-blocks of Life: Part 1

In this section we cover:

1) Particles
2) The basics of electron orbitals
3) Atoms
4) Ions

The processes of life are comprised of physical materials that obey the laws of physics. Here we discuss what those basic materials are, how they look and some simple characteristics that result from how they come together.

The Particle

The tiniest pieces of the Universe, as far as we can tell, are particles. They are incredibly small things, smaller than anything else in existence (except, of course, my housemate's chances of doing the dishes).

To give you an idea of their size: a *tiny* shred of paper on a table has an unfathomable number of particles in it. It *easily* has more particles than the number of people on Earth. You can reasonably think of particles as being *almost* infinitesimally small dots that have *essentially* no volume at all, approximated by theoretical things we call **point masses**.

There are many types of particles in the Universe, but the most important ones in the processes of life are the **proton**, the **neutron**, the **electron** and the **photon**. Protons are positively charged, and electrons are negatively charged, whereas neutrons and photons have no charge. By charge, I'm referring to the same concept as you encounter with batteries, and at a basic level the physics behind batteries is the same as the physics behind the processes of life: plusses and minuses interacting. "Like" charges repel each other. Opposite charges attract. So, protons and electrons attract each other. As for photons, let's put off that discussion until a later book, except to say that photons make up light.

Protons and neutrons have roughly the same mass, whereas electrons are about 1000 times *lighter*. In fact, electrons have so very little mass that their behavior is unlike anything you would expect. Why would this be? In physics (don't panic) there is an important theory known as the **Uncertainty Principle**. Although the details of this topic are far outside the range of this book, it suffices to say that the Uncertainty Principle makes it practically impossible to pinpoint exactly where *anything* is, especially the tiniest entities in our Universe. In a way, it's a matter of practicality, for how do you capture something so ridiculously small? It's like catching one specific grain of dust in a category 5 hurricane. The darn thing could be anywhere! The best you can do is say that the grain of dust in question is most likely within a fifty square mile radius. The analogy isn't perfect, of course, but when applied to electrons, it becomes necessary to draw them as fuzzy "probability" clouds, where the clouds represent the most likely space where the electrons will be found. For the sake of simplicity, electrons are commonly drawn as dots within little balloons (which are usually attached to atoms and molecules – but we'll get to that later). Sometimes, when we don't care about the actual physical position of the electron, we draw them as just simple dots. Don't be deceived, however, for electrons are not just little dots. Their exact nature is still the subject of a great deal of theory. Here are important things to remember about the little buggers: 1) they are super tiny, 2) negatively charged and 3) impossible to pinpoint exactly, so we typically draw them as clouds or balloons. Here is a picture for your brain to gnaw on.

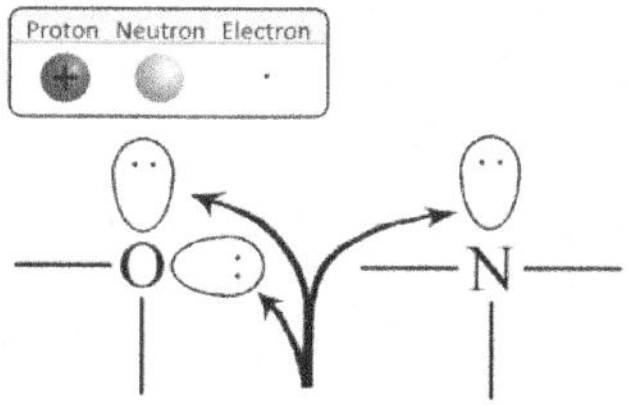

Sample electron clouds (two dots = two electrons in each). The "O" is an oxygen atom. The "N" is a nitrogen atom. Atoms are covered a little later.

Protons and neutrons clump together in various combinations to form the nuclei of atoms. Electrons, in turn, are clouds around the atomic nuclei, attracted by the positive charge. Most biology textbooks, (and many introductory chemistry books as well), portray electrons as "orbiting" around the nucleus in neat little circular paths. For reasons I explained above, this is not actually realistic, but for the purposes of most biological considerations, it works.

The Atom

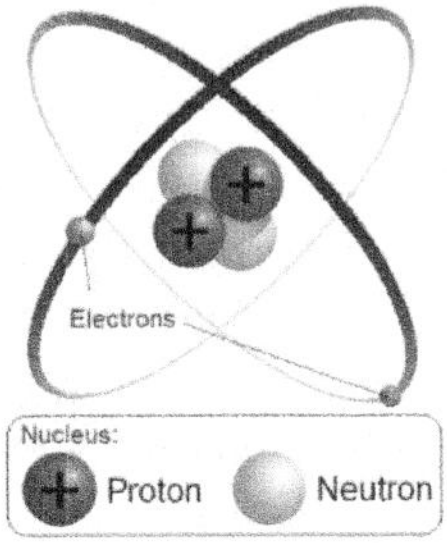

The number of protons that an atom has in its nucleus is called its **atomic number**, and it determines the type of **element** that the atom is. For example, carbon atoms have six protons in their nuclei. If you find an atom with six protons, it is carbon and never nitrogen, zinc, iron, etc. If you find a carbon atom, it *always* has six protons and never five or seven or some other number. The image of the atom to the right is a helium atom, because it has two protons in the middle – helium atoms never have more than two or less than two protons.

Unlike protons, the number of neutrons in atoms can vary, depending on the element, but the variation is usually not much. For example, most carbon atoms have six neutrons in their nuclei, while some have seven neutrons instead. Atoms that have the same number of protons but different numbers of neutrons are called **isotopes** of one another. So, the carbon with 6 neutrons is one isotope, while the carbon with seven neutrons is another isotope. The total mass of an atom is measured in **atomic mass units**[4] (amu). It is easily estimated by simply adding the number of protons and neutrons in the nucleus, because protons and neutrons each have a mass of about one amu. The mass of electrons is so incredibly small that we usually don't include them in the calculation. We indicate what isotope we are talking about by writing its atomic mass to the upper left side of the element's symbol:

^{12}C	^{13}C
Carbon 12	Carbon 13
6 protons	6 protons
6 neutrons	7 neutrons
mass: 12 amu	mass: 13 amu

[4] One amu equals approximately 1.66053886 x 10^{-24} grams. One amu = 0.0000000000000000000000166053886 grams. The amu is like one person living amongst 90,000,000,000,000 populated Earths.

Most atomic nuclei are *very* stable things, resisting all kinds of energy: heat, vibration, light, my ex-girlfriend's constant nagging, you name it. Electrons, on the other hand, can be ripped away from their orbits without much trouble – we'll get back to that. Some isotopes, however, are *not* stable and spontaneously eject various forms of radiation from the nucleus, without any external reason. These isotopes are said to be **radioactive**. When radiation emits from the nucleus, the atom changes isotope and atomic number. This is basically what happens to uranium and plutonium as they give off radiation to slowly transform into lead. For example, the first step in the breakdown of uranium-238 into lead is:

$$^{238}_{92}U \rightarrow\ ^{234}_{90}Th + {}^{4}_{2}He^{2+} + 2e^{-}$$

This reaction basically means that the uranium-238 nucleus boots out a quartet of two neutrons and two protons, written as ${}^{4}_{2}He^{2+}$. The nucleus left over is a thorium nucleus, which itself is unstable and eventually makes it own adjustments to become something else in time. Anyway, that's enough of that for the moment . . .

If you look at the periodic table you will see over one hundred different elements, each identified by their atomic number. Many of the elements with atomic number over ninety are radioactive. Since they are unstable, you cannot find much of them in Nature anymore. They've mostly broken down by now.[5] Anyway, there are 92 naturally occurring more-or-less stable elements, each having unique properties. Here is a table of the common elements that are found in living systems:

Element name	Atomic Symbol	Atomic Number
Hydrogen	H	1
Carbon	C	6
Nitrogen	N	7

[5] In case you're wondering, most, if not all, of these atoms came from stars. There's a lot of quantum mumbo-jumbo and mathematical hubbub, but basically stars are so incredibly hot that they smash atomic nuclei together, combining little nuclei into bigger nuclei. At some point in the distant past, stars emitted these atoms into space, which gradually formed together into planets . . . That or it all came out of a big galactic magic hat. I like to keep an open mind.

Oxygen	O	8
Sodium	Na	11
Phosphorus	P	15
Sulfur	S	16
Chlorine	Cl	17

Let's move on to discuss the electrons around the atom – those sketchy little cloudlike thingies. An atom will have an equal number of electrons around it as it has in its nucleus. So, a carbon atom will have six electrons around it, since it has six protons in its nucleus. The six protons in the middle will give the atom a plus six charge, while its six electrons contribute a negative six charge, making it neutral (+6 – 6 = 0). Atoms that have unequal ratios of electrons to protons will have an overall charge and are called **ions**. We'll talk more about those later.

As we said earlier, electrons gather in clouds around atoms. These clouds are called **orbitals**, even though they are *not* circular in shape (despite simplified pictures). Some orbitals are spherical in shape, while others have stranger shapes, such as bowties and clovers. Below is an image of a few sample electron clouds. The image is not comprehensive by any measure, for there are lots of other possible shapes available. The first two drawn are the ones that are encountered the most in basic biology.

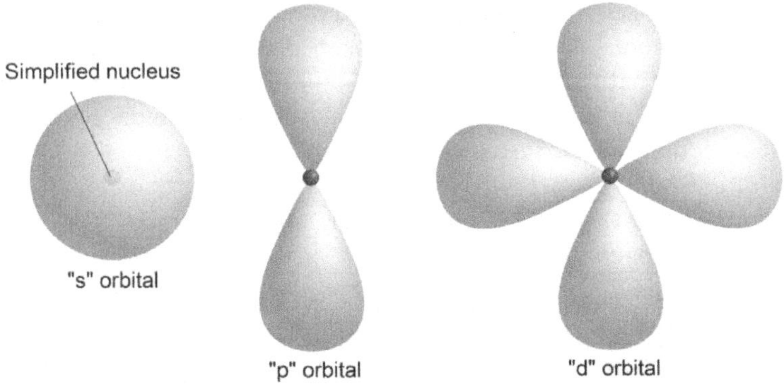

So what do these cloudlike orbitals represent? They represent the statistically most likely place where you would find an electron that is in that orbital. For example, consider the "p" orbital from the picture. The electrons in that orbital are most likely to be found somewhere within that bowtie shape at any given point in time. The bowtie squeezes down to zero in the middle, so that says the electron shouldn't be found there. However, the shape of the orbital is only a general guideline for where the electron could be found. *Technically* the electron can be found outside the orbital as well, but the odds are *overwhelming* that at any given moment it is somewhere *within* the bowtie.

I wouldn't worry too much if the electron orbital shapes seem alien. It is generally covered in detail in courses on chemistry, and it's not necessary to understand the concept in its fullness at this moment. Just be aware that the actual physical shape of the orbitals ultimately affects the physical shape of the atoms, and that ultimately affects how atoms interact and fit together to make molecules. Those molecules, in turn, have definite shapes that are critical in the reactions of life. The molecules inside you recognize and react with each other because of their characteristic shapes. So, by extension, the fact that electrons orbit around atoms in weird ways makes life possible.

Leaving their shapes behind, let's look at a few other characteristics of electron orbitals.

Each electron orbital can hold two electrons, which are said to be "paired." Electrons *like* to be paired[6], despite the fact that they both have negative charges. In fact, as a hard and fast rule, unpaired electrons tend to be rather unhappy, and their attempts to get paired leads to quite a bit of chemistry. It's a bit like dating in that sense. Don't get me wrong, molecules and compounds (and metals) *do* exist that have unpaired electrons, so the rule isn't perfect, but these compounds also tend to be unstable in the long run, breaking down in the direction that allows their electrons to become paired.

Multiple orbitals around atoms are divided into sets called **electron shells**. The first shell is closest to the nucleus and has only one orbital (which looks like a sphere – the "s" orbital in the picture above). Since the first shell has only one orbital, only two electrons

---

[6] It's actually due to a strange quantum mechanics principle known as the "spin number," as well as minimizing magnetic fields. I once tried to explain this to a student I was tutoring, and his head exploded, so I won't go into detail about spin numbers. All joking aside, just know that electrons like to be paired because of quantum mechanics.

can be in it. The second shell is a bit farther from the nucleus, is bigger and has four orbitals in it (one sphere and three bowties – one on each of the three-dimensional axes). Each of these orbitals can hold two electrons. So, with four orbitals in the second shell the total number of electrons that can fit there is eight. The third shell has nine orbitals in it (of a variety of weird shapes), which means the third shell can hold up to eighteen electrons. Most of the time, only four of the orbitals are filled, however, meaning that usually only eight electrons are found in the outermost shell of an atom in a stable form (the **octet rule** we see in chemistry classes).

There are technically an infinite number of possible shells: the higher the shell number, the higher the energy of the electrons that are in it. The lowest shells fill up first, though, so we rarely deal with the higher level shells. The row on the periodic table that the element is in determines the base number of shells it will have. For example, hydrogen is in the first row, so it has one shell. Carbon is in the second row, so it has two shells. Chlorine is in the third row, so it has three shells. Most biological chemicals are made from elements in the first three rows, so understanding of the first three shells will suffice to allow you to study the vast majority of topics in biology. These shells are summarized here:

Electron shell #	No. of orbitals	Max electrons
1	1	2
2	4	8
3	9	18

An atom's outermost shell is called the **valence shell**. Atoms generally interact with each other exclusively through their valence shells. That, in essence, is where chemistry happens. If we ignore the actual shape of the individual orbits, we can use the popular strategy of depicting the shells as just circles. The next image shows a few sample atoms with their electron shells. Some are happy. Some are not. Can you figure out why?

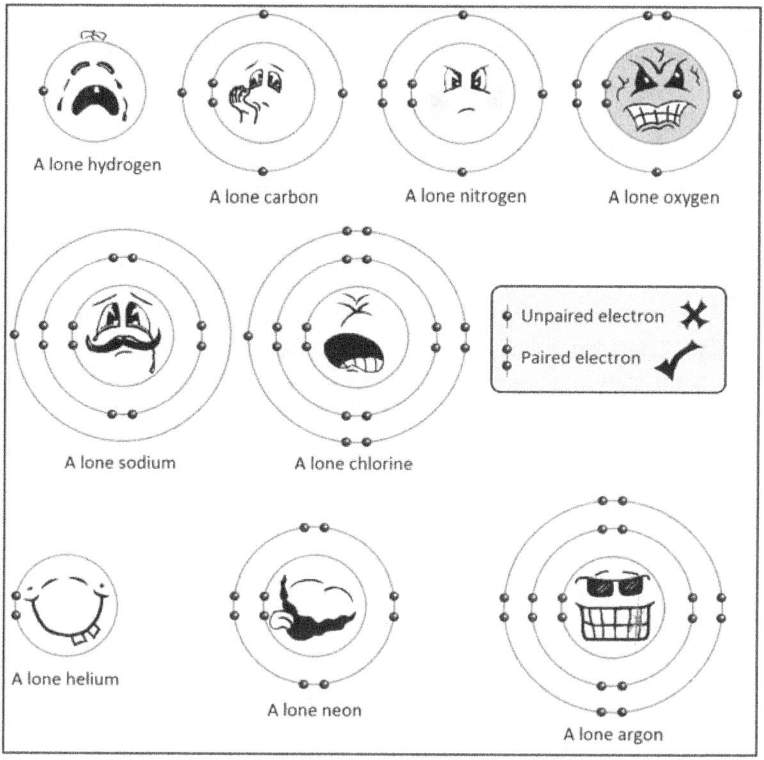

This image shows lone atoms. The hydrogen and helium have only one electron shell. Carbon, nitrogen, oxygen and neon have two shells, while sodium, chlorine and argon have three shells. Notice that the first two shells of sodium, chlorine, neon and argon have the maximum electrons in them (two and eight, respectively). The third shell of argon has eight electrons (which is typical, even though the third shell can technically have as many as eighteen). So why are some happy and some sad? Here is a golden rule of chemistry: **Atoms are happiest when their valence shells are full**. The reasons are, as always, due to quantum mechanics, but you can think of it like this: if the shells are not full, the atom will be lopsided in one way or another.[7] Nature doesn't like that. Further, since electrons like being paired, the electrons in the full valence shells are all paired off.

The pattern is simple:

2, 8, 8 make atoms great.

Helium, neon, and argon already have full outer shells. These atoms are completely happy by themselves. They don't need to worry about anything and rarely interact chemically with the world around them. Since they don't interact much with other atoms, (and are typically found as gases), we call them the **noble gases**.

Look at the other atoms in the picture. Their outer shells are not full. They are unhappy, unstable and lopsided. How can they fill their outer shells? One option, of course, is to simply add or subtract electrons. This makes an ion, and some atoms do indeed take this pathway. Here's what hydrogen, sodium and chlorine can do, (leaving carbon and nitrogen for the moment):

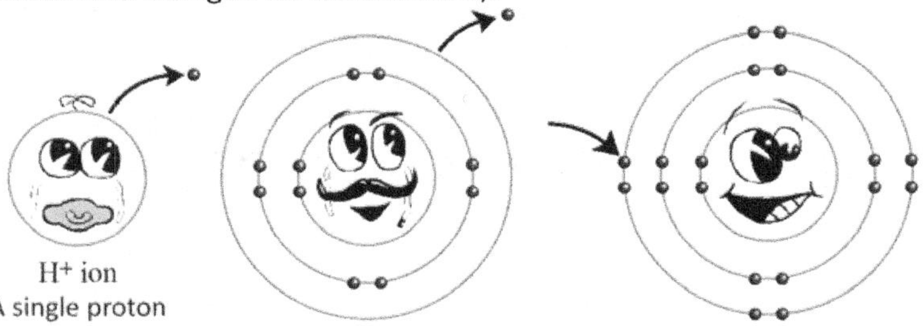

H^+ ion
A single proton missing an electron.

Na^+ ion
It is missing an electron from its outer shell.

Cl^- ion
It has an extra electron in its outer shell.

In the above image, we see three different ions: H^+, Cl^- and Na^+. The first one might surprise you. Hydrogen is probably the only atom that has the luxury of just getting rid of its electron altogether – it can become a simple proton, a wanderer, a vagrant, living carefree without the worries of all this orbital business. That must be nice. It's not necessarily the best option, and it doesn't always do this though, as we will see, for most hydrogen atoms would rather bond to another atom. The sodium takes a similar strategy of abandoning an electron.

[7] It will be lopsided not just physically, but magnetically as well. Also, the third shell doesn't need 18 electrons to avoid being lopsided. Eight electrons will suffice. For the purpose of discussion we will consider 8 electrons as making the third shell "full."

By doing so, the shell below it now becomes its valence shell, full and *not* lopsided. The chlorine simply gains an electron from somewhere (perhaps an atom, like sodium, that is stabilized by losing an electron) and becomes stabilized for the same reason.

The important thing to understand about these atoms is that they now have an imbalance in their proton to electron ratio. They are ions. The new hydrogen has one proton but no electrons, so it has a positive charge. The new chlorine has seventeen protons and eighteen electrons. That means it has a -1 overall charge. The new sodium has eleven protons and ten electrons. That gives it an overall charge of +1.

Many atoms form ions in this fashion. Here is a quick list of those that are relevant to biology: K^+, Ca^{2+}, Fe^{2+}, Fe^{3+}, Mg^{2+} and Zn^{2+}. The "2+" above an ion, (such as for calcium Ca), means that it is missing two electrons. Also, not all ions are made from single atoms. Some ions are actually groups of connected atoms that collectively have extra or missing electron(s). These are called **polyatomic ions**. Dihydrogenphosphate ($H_2PO_4^-$), ammonium (NH_4^+) and bicarbonate (HCO_3^-) are examples. You'll see many of these over and over, so don't sweat what they are at the moment.

Getting back to nitrogen and carbon, not all atoms can pull off this trick of gaining or losing electrons. So, don't think of the ion as the default cure for getting full electron shells. Atoms such as nitrogen and carbon generally do not use this strategy in stable compounds because it results in too much charge built up in too little space. They must therefore interact with other atoms through bonds in order to get full valence shells. This forms molecules. Although hydrogen has the option of abandoning its electron, that strategy is the exception not the rule. Most hydrogen atoms also participate in bonds as a result of the need to have a full valence shell. Bond and molecule formation will be the focus of the next unit.

Summary

In this section, we discussed the basic particles that are important in biology and how they come together to make atoms.

The basic particles are:

- Proton – positively charged mass found in the nucleus of atoms.
- Neutron – neutral mass found in the nucleus of atoms.
- Electron – negatively charged smaller mass that "orbits" around the nucleus in funky shaped clouds called orbitals.
- Photon – neutral particles that make up light.

Here are the basics we covered about atoms:

- Positively charged nucleus has protons and neutrons in it. Negatively charged electrons "orbit" in funky shaped probability clouds around the nucleus. Each orbital can hold two electrons.
- The number of protons in the nucleus is the atomic number and determines what kind of element the atom is.
- The nuclei of elements in the 90's and above tend to be unstable and break down by radiation, spitting pieces out of the nucleus to make a smaller nucleus, repeating the process until the nucleus is finally stable. The atomic number (equal to number of protons in the nucleus) changes when this happens.
- Electron orbitals are divided into sets called shells. The first shell has only one orbital (an s orbital), shaped like a sphere. The first shell can hold only two electrons. The second shell has four orbitals of various shapes. The second shell holds up to eight electrons. The third shell has nine orbitals and can hold up to eighteen electrons, but most atoms are happy having eight in the outer shell. There are more than three shells, but the first three are the most relevant to the study of biology. The outermost shell is called the valence shell and it's where chemistry between atoms takes place.
- Electrons like to be paired and atoms don't like to be lopsided, so atoms prefer to have their valence shell full of paired electrons. 2, 8, 8 makes atoms great. Hydrogen and

helium have only the first shell, so they only need two electrons to have a full valence shell. All others have two or more electron shells, so they need at least eight electrons in their outer shell. Elements in the third row of the periodic table (and below) can have more than eight electrons in their outer shell, but most atoms are happy having just eight.

- Some atoms acquire full outer shells by gaining or losing an electron. These atoms have an imbalance in their proton to electron ratio. So, they are charged. We call them ions. The word ion can also refer to molecules that have imbalances in proton to electron ratio. We call these molecules polyatomic ions. Atoms that are unable to make full valence shells by becoming ions must interact with other atoms to become molecules.

Unit 4

Nonliving Building-blocks of Life: Part 2

In this section we cover:

1) Atomic bonding – nonpolar and polar
2) Basic molecular formation and shape
3) Electron lone pairs
4) Hydrogen bonding
5) Polar molecules

Atomic Bonding

Atoms tend to be most stable when their outer shells are full and each electron is paired. Once again, the third shell can actually hold up to eighteen, but most atoms are happy with eight. Atoms that have full valence shells due to their inherent number of electrons (the hermit-like helium column in the periodic table) are **inert**. Unfortunately, most of the 'unhappy' atoms in other parts of the periodic table cannot simply add or subtract spare electrons to fill their shells (like chlorine and sodium), because most of them cannot handle the charge that results. For example, an isolated carbon with eight electrons in its outer shell would have an overall –4 charge all packed within one atom's space. Electrons like pairing, but when too much negative charge piles up in one place the repulsion of all the charge starts to overpower the electron pairing, breaking up the bunch. Similarly, too much positive charge tends to attract electrons back to the positively charged atom. So, what is an atom to do when it wants more electrons but cannot do it on its own? The answer is to share. Atoms share electrons through bonds with other atoms that similarly want their valence shell plump-full of paired electrons. When atoms share electrons in this way, they are said to form a **bond** between them. This type of union is called a **covalent bond**, and the formation and breaking of such bonds are called **chemical reactions**. The existence of two or more atoms bound together through covalent bonds makes up a **molecule**, a concept we will return to shortly. For the moment, a simple covalent bond would look like two atomic nuclei sharing electron clouds:

3D image of a hydrogen-hydrogen bond

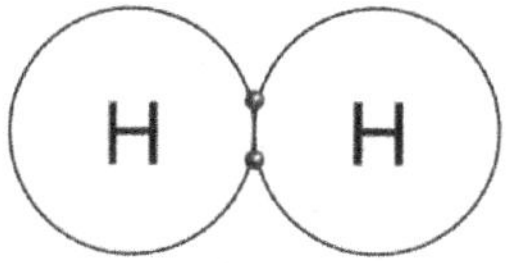

Simplified valence shell image of a hydrogen-hydrogen bond

H—H

The lazy chemist's shorthand stick verson of a hydrogen-hydrogen bond. Crude by affective.

The more electrons that are shared between atoms, the stronger the overall bond will be. Covalent bonds found in living systems can have as many as six electrons shared between neighboring atoms, but two and four are the most common. Neighboring atoms that share two electrons (like the H-H in the image above) are said to have a **single bond**. Sharing four electrons is called a **double bond**. Sharing six electrons is called a **triple bond**. These are the strongest. Sharing more electrons than this is a phenomenon that only takes place in atoms not typically seen in living systems.

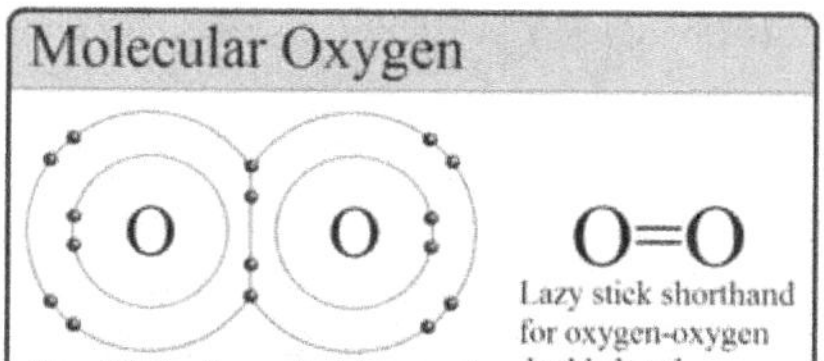

Simplified valence shell image of an oxygen-oxygen double bond.

Lazy stick shorthand for oxygen-oxygen double bond.

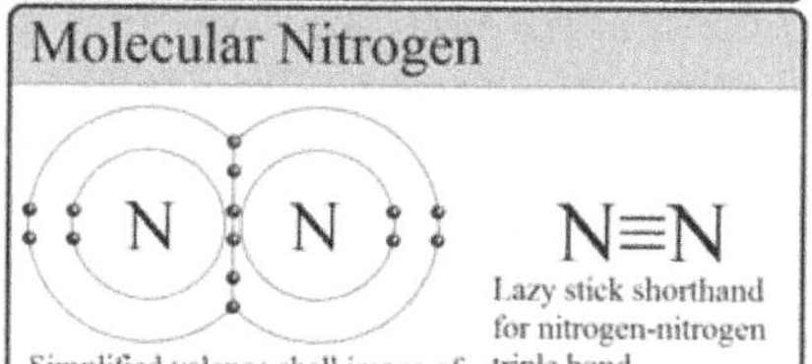

Simplified valence shell image of a nitrogen-nitrogen triple bond.

Lazy stick shorthand for nitrogen-nitrogen triple bond.

When atoms share electrons, they don't necessarily share them fairly. Some atoms tend to hog the electrons more than others. When this happens, you get a bond that is said to be **polar**. In a polar bond, the shared electrons are pulled closer to one atom than to the other. This typically happens when the atoms that are sharing electrons are not of the same element, such as oxygen bonding with carbon. The elements in the top right part of the periodic table (excluding the weird hermits in the far-right "helium" column) are more selfish than the other elements, a characteristic called **electronegativity**. The more electronegative an element is, the more it will tend to try to hog electrons. Curiously, carbon and hydrogen have almost identical electronegativities, and so when these two elements get together in bonds, they share electrons equally. When electrons are shared equally, you get a **nonpolar** bond. All the bonds in the images on the previous page are nonpolar bonds, because they involve elements of the same type bonding with each other. For example, even though oxygen likes to hog electrons, when two oxygen atoms are bound together, they try to hog the electrons equally, cancelling the effect entirely.

Let's take a closer look at polar bonds. Polar bonds look like electron clouds that are not positioned evenly over the atoms: one atom will have more of the electron cloud over it, while the other will have less. Since electrons are negatively charged, the more-covered atom will have a sort-of negative charge (a "partial negative," represented by the funny little symbol δ^-), while the other less-covered atom will have a sort-of positive charge (a "partial positive," δ^+). Overall, the two-atom combination is still neutral, because there are an equal number of electrons to protons, but in this situation, the electrons are not evenly distributed, so the effect is that one end has a negatively charged pole, while the other end has a positively charged pole.

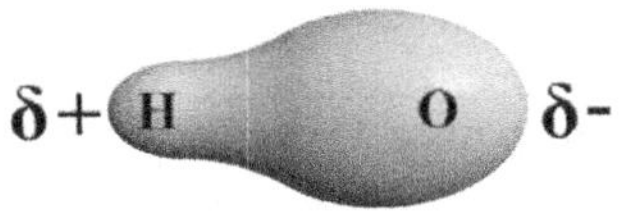

This last image is a hydrogen-oxygen bond – a *very* important polar bond in biology – which for some reason came out looking more like a bowling pin or one of those mutated peanuts. I hate those things, because the little end always has some gnarly looking runt in it. Anyway, the main thing to understand about polar bonds is that it causes neighboring polar molecules to get attracted. They tend to want to line up in various geometries. For example, H-Cl also looks like the mutated peanut image above, and in really *cold* samples of H-Cl,[8] they tend to line up as shown so that the positive poles attract the negative poles:

[8] Heat makes molecules move around a lot, so the effect wouldn't be very noticeable in HCl unless the sample is really cold. The point is still the same though.

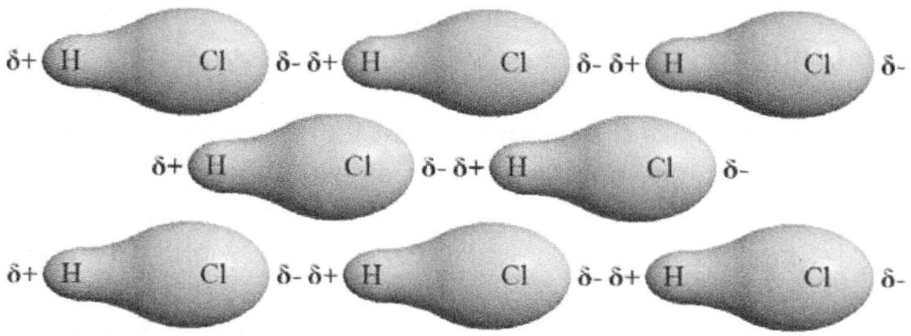

In the previous unit about atoms, we talked about ions. A simplified way of looking at ions is to consider them to be extreme examples of polar bonds. Instead of the electron clouds being off-center, the electron is ripped away from one atom entirely and hovers around the other atom instead, creating a positively-charged ion and a negative-charged ion. The two ions are then held closely to one another by the attraction of opposite charges.

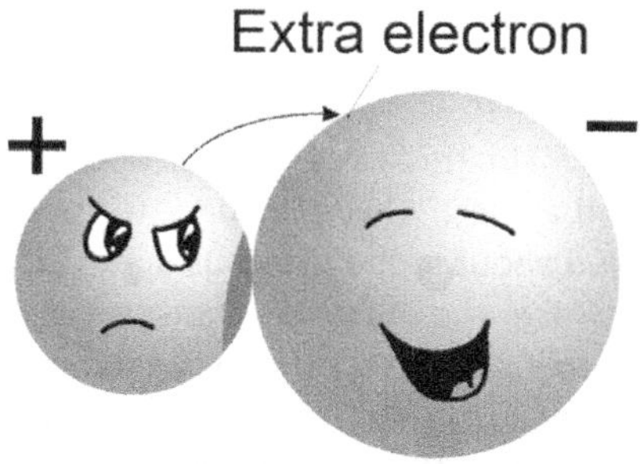

The attractive force that holds ions together is called an **ionic bond**. This occurs, for example, when one atom only needs one electron to fill its valence shell with paired electrons, whereas the other can have a valence shell full of paired electrons if one electron is *removed* from its outermost shell. So, one atom gives up its electron and acquires a +1 charge, while the other atom gets the electron and subsequently has a -1 charge. Sodium chloride (i.e. Na^+ and Cl^-), also known as table salt, is an example of this.

The Molecule

Molecules are collections of atoms that are covalently bonded with each other. The number of bonds that an atom will engage in is determined by the number of electrons it wants to pair up in its outer valence shell. Hydrogen has only one electron, but its valence shell has room for two, so it participates in simple, single bonds. Oxygen has room for two more electrons in its outer shell, so it engages in two bonds. This may come in the form of two separate single bonds, or one double bond. Carbon has four electrons in its outer shell, but would like to have four more, so it participates in four bonds. It might make these four bonds as four single bonds, or as two single bonds and one double bond, or as two double bonds, or one triple bond and one single bond. Nitrogen needs three electrons to fill its shell so it participates in three bonds, and as with oxygen and carbon, these bonds can be through any combination of single, double and triple bonds. The image to the left shows some example molecules.

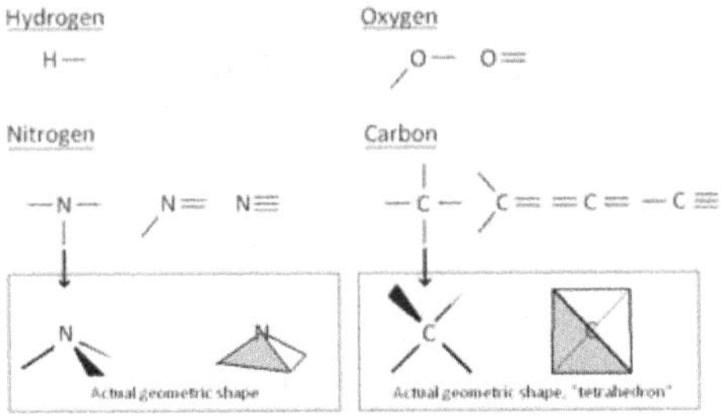

In this image you'll notice that some of the bonds are at weird angles. This goes back to the funky electron orbital shapes we talked about in the previous unit. That's why I made such a fuss about it. Take the nitrogen that participates in three single bonds for example. It forms a shape like a trigonal pyramid, with the nitrogen at the top. The carbons that participate in four single bonds form a shape called the **tetrahedron**, a four-sided geometric thingamajig with triangular sides. The carbon is deep in the center of the shape. While this gets a little too far afield for the present study, it should definitely be kept in mind that the true shape of the molecules differ from the way they are often drawn on paper.

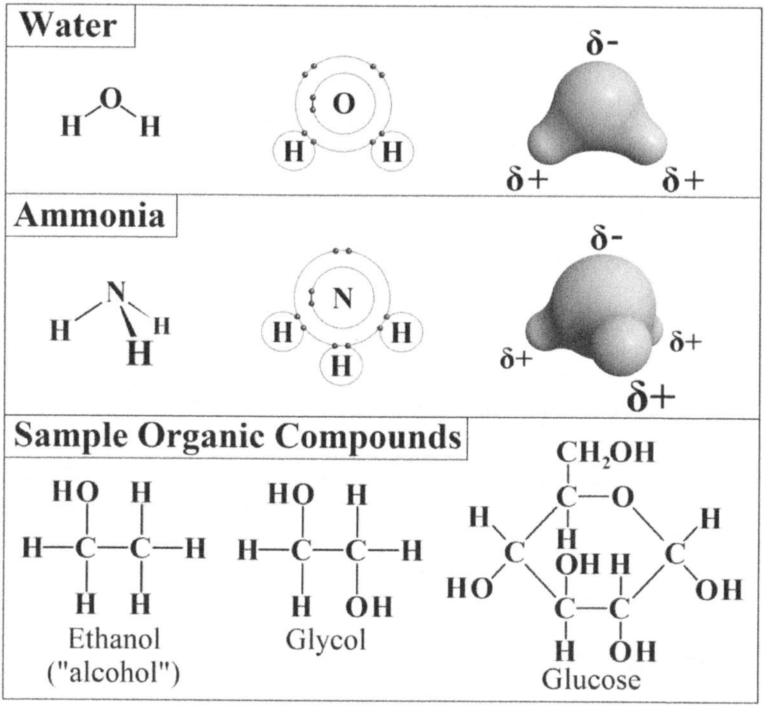

No matter the weird shapes, since carbon atoms can have up to four bonds, they can connect a lot of atoms together. Because of this, carbon is often found as the backbone for biological compounds. These molecules are called **organic molecules** and are the basis for the incomprehensible study of organic chemistry. They exist in Nature because 1) they can, and 2) to terrorize chemistry students the world-over. Organic molecules, along with water (H_2O), ammonia (NH_3) and a few others form the primary ingredients in the processes of life. The image above is of water, ammonia and a few sample organic molecules. Don't worry about the details of the organic molecules shown. We will encounter them again as we need them. They are here just to give you an idea of what they would look like and how they are drawn.

Lone Pairs

Electrons within the outer shell of atoms that are already paired do not participate in bonding and instead gather in bulges that stick out from the atom (not shown in previous images for simplicity).

These are called **lone pairs** and are very characteristic of atoms like oxygen and nitrogen.

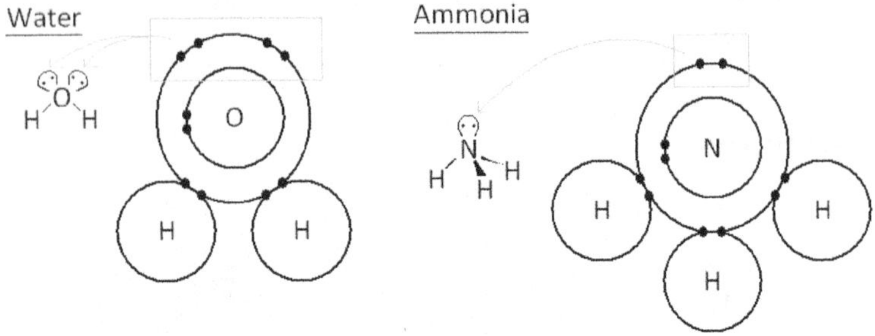

The fact that lone pairs exist in atoms like oxygen and nitrogen is very important. The main consequence is that they represent negatively charged clouds poking out the side of atoms. This means that if any other positively charged thing comes nearby, it will tend to be attracted to the lone pair. One VERY important example of this is the hydrogen bond.

The Hydrogen Bond

As we saw earlier, hydrogen-oxygen bonds are very polar because the electrons in the bond are hogged by the oxygen. As a result, hydrogen tends to have a partial positive charge while oxygen gets a partial negative charge. The hydrogen's partial positive charge makes it tend to stick to negatively charged objects in the immediate environment. Further, the oxygen has two rabbit-ear-like lone pairs bulging out from its side, which are both very negatively charged. These act to attract other hydrogen from nearby hydrogen-oxygen bonds, creating a network of sticky molecules. Hydrogen-nitrogen bonds behave in the same way. As we will see in future sections, this unique "bond" gives water and many other biological molecules very unique characteristics that are crucial for the processes of life.

Polar Molecules

When we discussed chemical bonds, we brought up the topic of polar covalent bonds, an example of which you just saw in the water molecule. The concept of polarity can be extended to molecules at large. If molecules have polar bonds that cause the overall charge distribution in the molecule to be unevenly spread out, the molecule itself is said to be polarized.

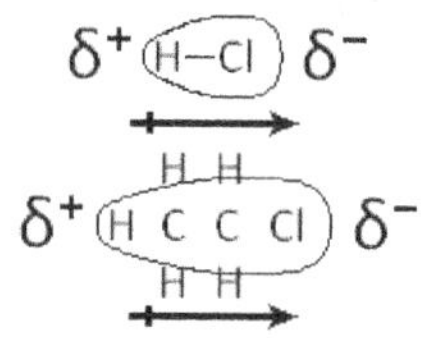

Simplistic way to think about polar molecules, ignoring the details about the individual bonds.

One important aspect of this topic is that polar molecules do not mix well with non-polar molecules. Why is this so? Well, polar molecules tend to stick to each other because their partially charged ends attract each other. In the attempt to stick to each other, they push nonpolar molecules out of the way. It's not that polar molecules dislike nonpolar molecules, they just like polar molecules *much* more.

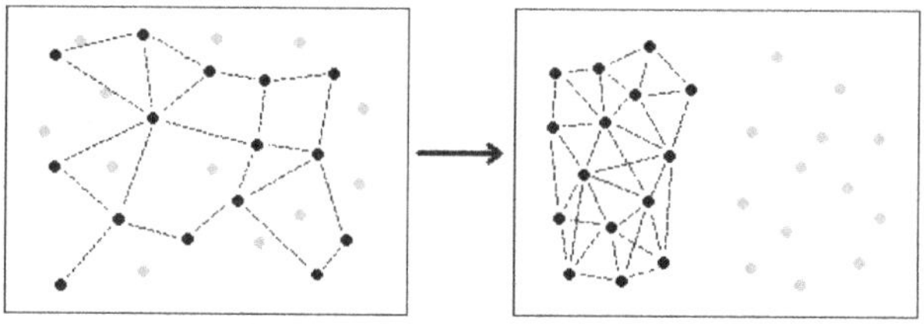

• Polar molecules
• Nonpolar molecules

The polar molecules attract each other, unintentionally excluding the nonpolar molecules, causing them to separate.

Since water is polar it mixes well with other polar molecules. We call molecules that mix well with water **hydrophilic** ("water loving"). Compounds that are non-polar, like oil and wax, do not mix well in water. These we call **hydrophobic** ("water fearing"). This concept will be important in future studies about cell membranes.

With that, we bring our discussion of the building blocks of life to a close. In the next section we will look closer at water and the qualities that make it important for the existence of life.

Summary

In this section, we covered the basics of atomic bonding.

- Atoms share electrons in an effort to fill their valence shell with paired electrons. The bond that forms is called a covalent bond. The collection of atoms participating in covalent bonds is called a molecule.
- Since electrons like pairing, bonds almost always involve electron pairs. A single bond has two paired electrons in it. A double bond has two sets of paired electrons (four electrons total). A triple bond has three pairs of electrons in it (six electrons total).
- Elements in the upper-right part of the periodic table, (excluding the "helium" column, noble gases), are more electronegative than the rest. These atoms hog electron, so bonds they make with atoms in the other parts of the periodic table are polar.
- Polar bonds involve unevenly shared electrons in a covalent bond and cause the participating atoms to have partial electronic charges, δ^- and δ^+. These partial charges attract other partial and/or full charges from the environment, affecting how nearby molecules behave.
- An ionic bond is an extreme form of a polar bond, where the electron from one atom is ripped away and goes to the other atom entirely. This makes an ionic bond.
- Carbon can participate in four bonds total. It is commonly strung together to make the backbone of big molecules.
- Nitrogen can participate in three bonds total.
- Oxygen can participate in two bonds total.
- Hydrogen can only participate in one bond.
- Covalent bonds are usually at strange angles, so that the molecules are difficult to draw on paper. These strange bond angles are due to quantum mechanics. The bond angles give molecules specific shapes.
- Electrons that are paired prior to covalent bonding form "lone pairs," sticking out of the side of the atom. These are negatively charged balloons that attract partial and full positive charges nearby.

- Hydrogen bonds involve the partial positive charge of the hydrogen bonded to either nitrogen or oxygen (H-N, H-O) with the lone pairs of oxygen or nitrogen. It is a very important bond in biology.
- Polar molecules have uneven distribution of electron clouds over the entire collections of atoms. This causes the entire atom to have regions of partial negative and partial positive charge. Nonpolar molecules have electron clouds that are evenly spread over the entire collection. So, they do not have partial charges. Polar molecules attract each other, excluding nonpolar molecules. This is why they do not mix. Water is polar. Oils are nonpolar.

Unit 5

Water and Life

In this section we cover:

1) The benefits of water's freezing characteristics
2) The benefits of water's solvent ability
3) The benefits of water's temperature moderating ability.

Water's Hydrogen Bonds

In the previous unit, we discussed the nature of lone pair electrons and how they bulge out of the sides of atoms. Each lone pair forms a cloud of negative charge to which nearby positive things are attracted. Let's quickly rehash the example of water. The image to the right is a water molecule.

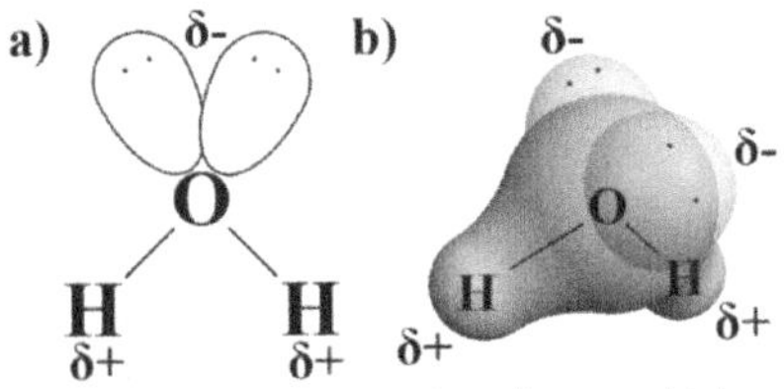

a) Simple stick representation of water with lone electron pairs. b) 3-D water molecule with lone pairs shown (omitted in previous 3-D image of water). Notice that it is similar to carbon's tetrahedral bonding structure.

The image shows the two lone pairs as well as the two polar bonds. Taken together, the entire structure is tetrahedral, just like carbon when it has four single bonds. Each of the hydrogen atoms in water has its electron drawn toward the oxygen, so that H carries a partial positive charge. These positive charges get attracted to nearby lone pairs from other water molecules to form **hydrogen bonds**. Although these are called "bonds," they can be broken and reformed much more easily than ionic or covalent bonds. At room temperature, for example, water molecules are moving around sufficiently quickly that hydrogen bonds are constantly being formed and broken. When the temperature drops, so does the rate at which the molecules tumble. The slower movement allows the water molecules to begin to line up based on their hydrogen bonding. The water slowly expands in volume and gradually freezes. When this happens, the water (now ice) becomes less dense and floats to the surface.

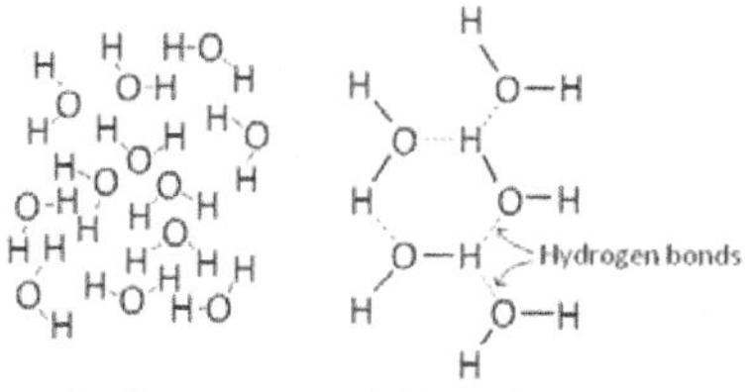

Liquid water   Solid water (ice)

Since ice floats, it forms a thin barrier at the top of lakes and streams, helping to trap in the remaining warmth underneath. This slows the freezing of the water at the bottom and prevents fish and other organisms from freezing during the winter. Otherwise, the population of fish and other critters in the inland water systems would be very small (if not zero). Since a lot of land ecology depends on fish for nutrients, life as we know it would probably be very different if water froze from the bottom up. Keep in mind that fish have been a major food source for mankind throughout its history, so without it who knows where we would be?

Water is a Good Solvent

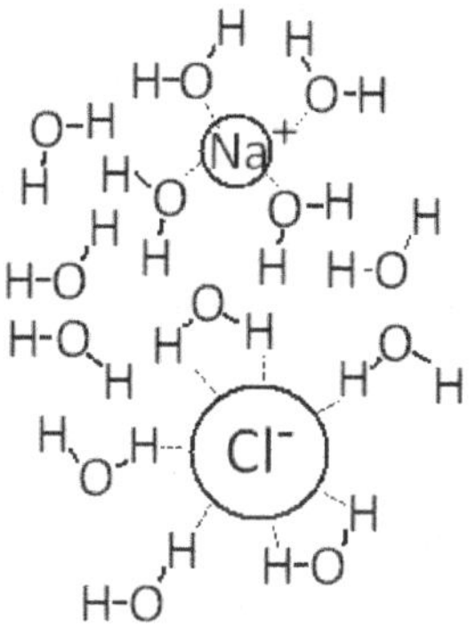

Salt dissolved in liquid water

Water is an excellent **solvent**. Solvents are generally liquids that dissolve other molecules. So, for example, if you take a small spoonful of table salt (Na^+Cl^-) and pour it into a cup of water, you will begin to notice the salt dissolve into the water. This is because water is polar, and the sodium chloride is made of positive and negative charges. These charges interact with the partial positive and negative charges of the water molecules. Each ion gets ripped away and carried off by a number of water molecules, until the ions are so spread out in solution that you can't see them anymore.

Water also dissolves polar molecules by the same mechanism. Since polar molecules have positive and negative ends, water is attracted to these ends and pulls these molecules away from the other molecules. Sugar, for example, is a polar compound (not an ionic compound like salt), but it dissolves in water like salt does.

The fact that water is such a good solvent is important because it allows many of the molecules that are involved in the reactions of life to be dissolved so they can float around and find each other for reaction to do important biological things. Otherwise, they would just sit there, like the salt in the spoon, or like my housemate on the couch. Come to think of it, he would also do well to be introduced to the concept of water.

Water as a Temperature Moderator

You may have noticed that on incredibly hot days, rocks and other solids tend to heat up rather quickly. Water, on the other hand, heats up much slower. Why is this? Temperature is basically a rough estimate of how quickly the molecules in a material are moving around and/or vibrating. The hotter it is the more vigorously they move about and/or vibrate. In liquids, this means they tumble and roll past each other, continuing on through the crowd of similarly moving molecules. In solids, it means that the atoms are vibrating energetically, but are relatively locked in place. Water heats up slowly because the energy that it takes to get them moving, rolling and tumbling is high. This is because it takes energy to disrupt all those hydrogen bonds that are in water. The net result is that water tends to maintain relatively stable temperatures while the surrounding land gets hot or cold quickly. This is one reason that people enjoy living near large bodies of water; the slow change of water temperature helps keep the surrounding land cooler in the summer and warmer in the winter versus areas with little water. Deserts, for example, are quite low in water, and they are often extremely hot during the day, but quite cold at night. Living organisms have a lot of water on the inside, and this helps to keep their bodies in relatively constant temperature despite fluctuations in the temperatures outside.

Summary

In this section, we covered the importance of water in biological systems.

- Its hydrogen bonding makes it expand when frozen, forming a layer that protects the underwater communities from freezing in the winter.
- Hydrogen bonding makes water a very good solvent. It is easy to dissolve many types of salts and polar molecules in water. A lot of biological molecules are polar, so water helps to keep them floating around, maximizing their ability to interact with other biological molecules. This maximization speeds the rate of biological reactions, which would otherwise be too slow to support life.
- Water is an excellent temperature mediator and helps to keep our body temperatures relatively constant despite fluctuations in outside temperature.

Unit 6

Acids and Bases

In this section we cover:

1) Acids and Bases
2) The concept of pH and the pH scale
3) Acid/Base buffers inside living systems

As we have seen, water is made of one oxygen atom and two hydrogen atoms. The bonds between these atoms are very strong, but every once in a rare while, a hydrogen snaps off. The stray hydrogen, (which is literally just a proton, H^+), floats away but leaves its electron behind with the ^-OH. This remaining part is called **hydroxide** (^-OH). Here's a simplified view:

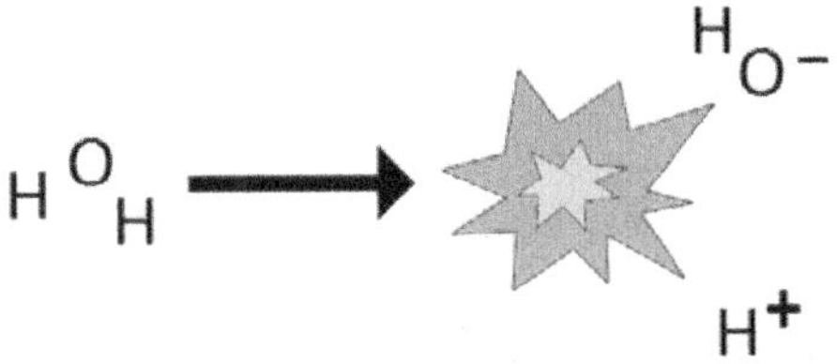

Water isn't the only molecule that can do this, though. In fact, a *lot* of chemicals are capable of breaking into pieces, releasing H^+. Molecules that release H^+ are called **acids**. HCl is an excellent example. When you put it in water, essentially every molecule breaks in half (whereas only a tiny fraction of water does).

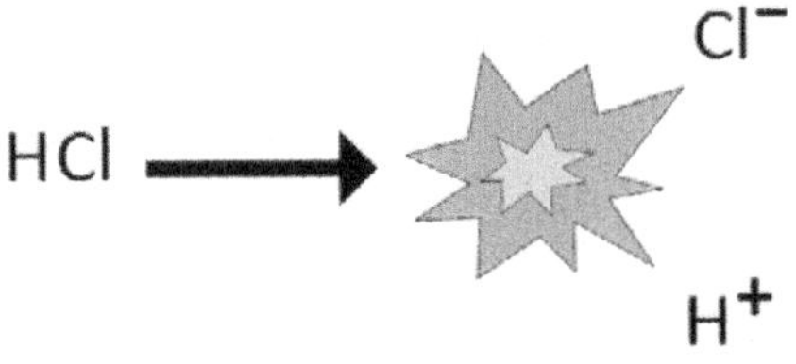

Since HCl breaks apart so easily and releases lots of H^+, it is a very strong acid.

Conversely, any molecule that breaks apart to release OH^- is called a **base**. NaOH is an excellent example. Like HCl, when you put it into water, it practically all breaks apart (dissociates) into ions:

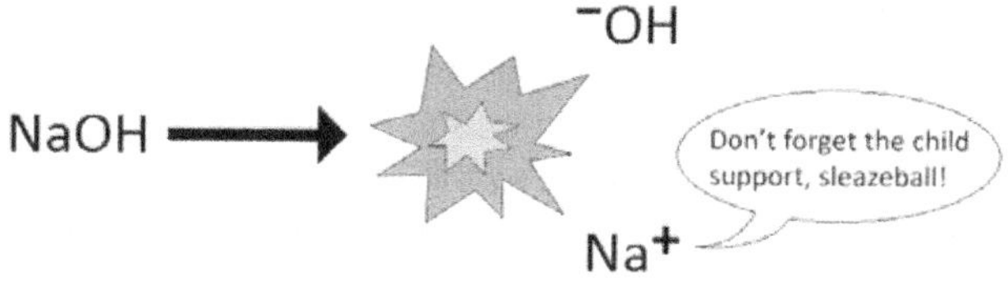

Since NaOH releases a lot of OH^-, we call it a very strong base.

If a solution of water is pure, there will be equal numbers of H^+ and OH^-. However, what if we put something else in the water that easily gives up H^+? Suppose we put HCl into the water. When this happens, the concentration of H^+ far outweighs the concentration of OH^-, not simply because H^+ is being added to the solution, but also because the H^+ rejoins with spare OH^- to recreate water. Water has a peculiar nature, however, for if you know the concentration of H^+, you can figure out the OH^- concentration by using the following equation:

$$[H^+][OH^-] = 10^{-14} \qquad (1)$$

The brackets "[ ]" around a chemical name means the "concentration of" and is typically measured in units of molarity (M).[9] So, from this equation, we can gather that as the concentration of H^+ increases, the concentration of OH^- drops, and vice versa. When there is more H^+ than OH^-, we say the solution is acidic. When there is more OH^- than H^+, we say the solution is basic or **alkaline**.

The way we measure the acidity or basicity of a solution is by the **pH** scale. That is determined by knowing the concentration of H^+ and using the following formula:

$$pH = -\log[H^+] \qquad (2)$$

---

[9] "M" stands for "molar," which is short for "moles per liter." Moles are a unit of quantity. One mole equals 6.22×10^{23} molecules. So 1 M, would be 6.22×10^{23} molecules in 1 liter of solution.

This formula demonstrates that solutions that have high H^+ concentrations have a low pH values. This is sometimes tricky to understand at first, so I've included a series of practice problems at the end of this section. It includes the answers as well. There is a shortcut you can use to approximate the pH of a system. If you are given the H^+ concentration in scientific notation (e.g. 1.234×10^{-5}), the pH will be approximately equal to the opposite value of the exponent (the little number above the 10). There is a little bit of error using this shortcut, but the actual pH will be between that guess value and the value one unit lower. So, here are some quick examples:

Example 1) $[H^+]$ 5.4×10^{-5} M   Guess: pH 4-5   Actual pH = 4.27

Example 2) $[H^+]$ 3.8×10^{-12} M   Guess: pH 11-12   Actual pH = 11.42

Example 3) $[H^+]$ 1.3×10^{-3} M   Guess: pH 2-3   Actual pH = 2.89

Example 4) $[H^+]$ 0.3 M   Guess: pH 0-1   Actual pH = 0.52

(Same as 3.0×10^{-1})

Be warned though! The only way this trick works is if the concentration is listed in scientific notation, specifically as $X.YZ \times 10^{-W}$, where X, Y, Z, and W are just digits. The units of measure *also have to be in M*. If there is something *other* than M after the concentration number, you have to do some mathemagic on the number to convert it. That is the focus of work done in chemistry classes. The odds are small you would need to do this in a typical introductory course in biology. At any rate, with a little practice you can get pretty good at eyeballing what the pH will be using this shortcut.

Anyway, **acids have low pH values** and **bases have high pH values**. Pure water has a pH of 7, and this value is considered **neutral**. Theoretically, the most acidic that a solution in water can get is 0. Conversely, the most basic that a solution in water can get is 14. The table below gives some examples of chemicals and their pH values.

pH	Example	
14	Drain cleaner	Most basic (alkaline)
13	Bleach	
12	Soapy water	
11	Ammonia	
10	Milk of Magnesia	
9	Baking Soda	
8	Sea Water	
7	Pure water	
6	Milk	
5	Vinegar water	
4	Tomato juice and beer	
3	Orange juice and grapefruit	
2	Soft drink (soda, pop)	
1	Stomach acid	
0	Battery acid	Most acidic

Humans and most other animals have an internal pH of around 7.3 to 7.4, but the internal pH of other organisms can vary greatly depending on the species. This value *cannot* change very much without having disastrous effects on the delicate living system. On the other hand, living organisms are chocked-full of all kinds of reactions that eject or soak up H^+. If these went unchecked, the pH in a living organism would fluctuate wildly (usually dropping, such as in muscle tissue during exercise), resulting in the untimely demise of the creature from the inside out. To counteract this, living systems have chemicals known as **buffers**. Buffers are just molecules and ions that act to soak up extra H^+ and OH^- floating around, keeping the H^+ concentration stable. Think of buffers as pH police, catching any stray H^+ that tries to get away. There are many important buffers in living systems, including HCO_3^- (bicarbonate), HPO_4^{2-} (mono-hydrogen phosphate) and many proteins and amino acids. We'll see many of them in future biology studies.

Practice pH Problems

1) What is the pH of a solution with a [H+] = 4.67 x 10^{-5} M?

Since the little number above the ten is "-5," we guess the pH will be between 4 and 5. Here is the calculation:

pH = -log(4.67 x 10^{-5}) = 4.33

In a graphing calculator:

1) (-) button
2) LOG button
3) 4.67
4) EE button (you may have to hit the 2ND button first)
5) (-) button then 5
4) ENTER button

In a simpler calculator, you'll have to experiment around to figure out how to get it to work. You often have to do the steps backwards though.

2) What is the pH of a solution with [H+] = 2.15 x 10^{-10} M?

Guess: pH 9-10.

pH = -log(2.15 x 10^{-10}) = 9.67

3) pH for [H+] = 1.334 x 10^{-2} M?

Guess: pH 1-2.

pH = -log(1.334 x 10^{-2}) = 1.87

4) pH for [H+] = 0.003244 M?

Guess: ? (If you convert this scientific notation, you'll get 3.244 x 10^{-3}, so we would guess the pH to be 2 to 3.)

The calculator doesn't care if it's scientific notation, so we type it in as is.

pH = -log(.003244) = 2.49

5) pH for [H+] = 9.88 x 10^{-11} M?

Guess: 10 to 11

pH = -log(9.88 x 10^{-11}) = 10.01

Summary

In this section, we covered the basic concepts of acids and bases.

- Acids produce H^+ in water (or take away OH^-).
- Bases give OH^- in water (or take away H^+).
- Acidic solutions have high H^+ concentrations and low OH^- concentrations.
- Basic solutions have high OH^- concentrations and low H^+ concentrations.
- Basic solutions are also called alkaline.
- The H^+ and OH^- concentrations in water are related by the equation $[H^+][OH^-] = 10^{-14}$. So, if you know the concentration of one, you can find the concentration of the other.
- The pH of a solution is given by the equation $-\log([H^+])$. So, if you know the H^+ concentration, you can find the pH.
- Acidic solutions have low pH (< 7).
- Basic solutions have high pH (>7).
- Fluctuating pH values are dangerous to living organisms, so they have buffers inside them that keep the H^+ concentration stable.

Unit 7

Some Simple Organic Chemistry

In this section we cover:

1) Functional groups in organic chemistry
2) Basics of organic chemistry stick models
3) The basic concept of metabolism
4) Monomers, polymers, dehydration synthesis, and hydrolysis

So far we've talked about particles, the atoms they make and how the resulting atoms come together to form molecules. Then we talked about water and looked at how it behaves in the context of acids and bases. Now let's return to our discussion of molecules, specifically those that have carbon in them (i.e., **organic molecules**). Carbon can have up to four single bonds, and in this situation, the carbon's bonds are in a four-sided pyramid-like shape called a **tetrahedron**. Carbon is especially good at connecting long chains of carbon-based molecules together. Ultimately, these molecules can have some complex overall shapes and structures if enough carbons are strung together. For example, here are a few of the shapes you'll encounter in a jaunt through the chemicals of life (see graphic, following page).

Cholesterol ("Lipids")

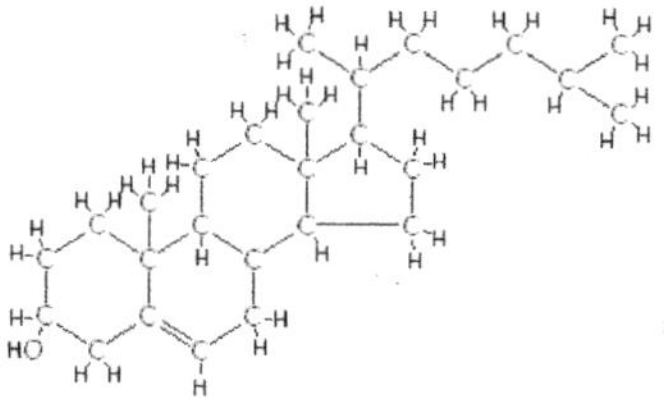

All details shown

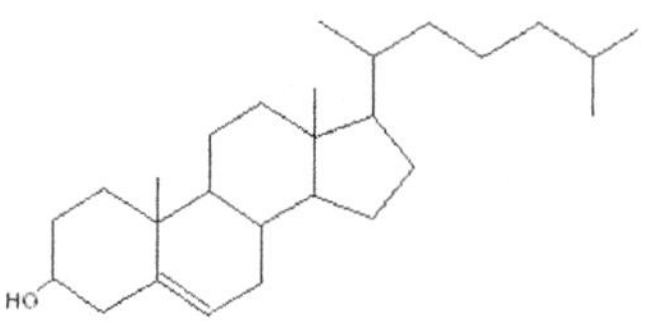

Simple shorthand stick notation

A Fatty Acid ("Lipids")

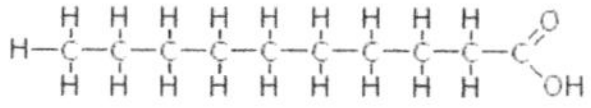

Detailed molecule

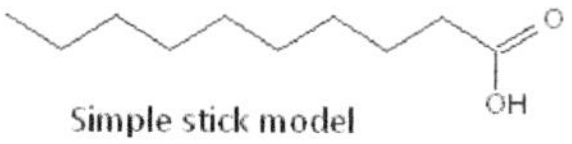

Simple stick model

Glucose ("Carbohydrates")

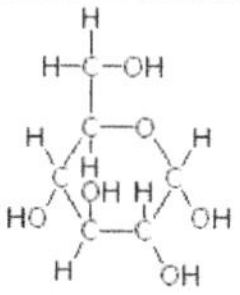

Detailed molecule

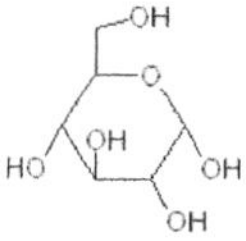

Simplified stick model

A Nucleotide, Adenosine ("Nucleic acids")

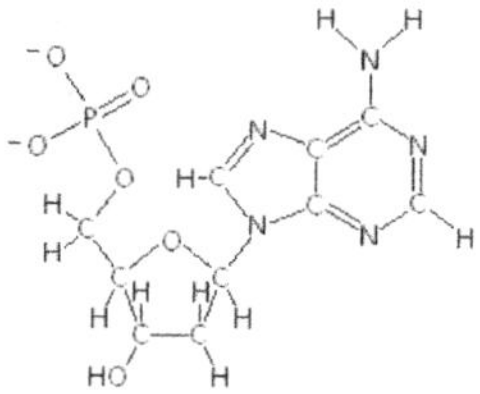

All details shown

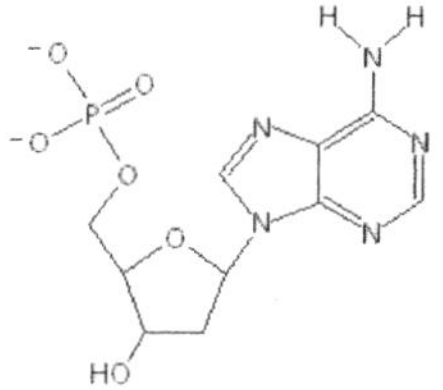

Simple stick model

Some Amino Acids ("Proteins")

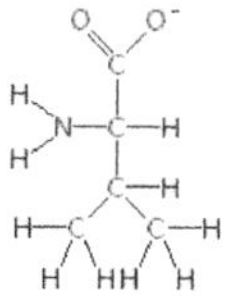

Valine

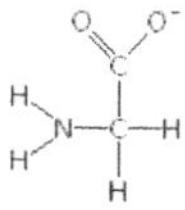

Glycine

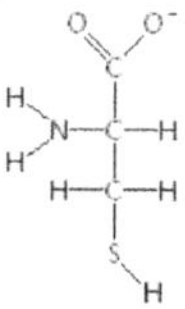

Cysteine

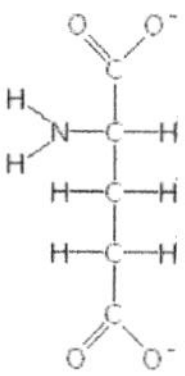

Glutamic acid

As you can see from the previous image, these molecules can get *incredibly* complex. Chemists are wild and crazy people though, so instead of drawing each "C" and "H" atom, they represent organic molecules as knobby little stick thingies. It's how they roll. The simplified stick models for some of the above compounds are included in the image. In these simplified models, the carbons are the intersections of lines, or corners if you will. An end of a line is also a carbon. All carbons have four bonds, and so any intersection that is not met by four lines is assumed to have enough hydrogen attached at that point to make up the remaining bonds. For example, an intersection with two lines is assumed to have two hydrogen attached. An intersection with three lines is assumed to have one hydrogen attached.

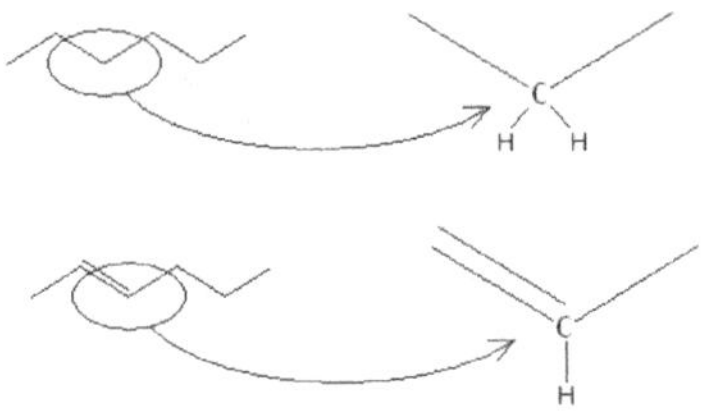

Stick models or not, carbon is the backbone of life, and from this backbone are attached an assortment of other atoms and groups of atoms that we call **functional groups**. Functional groups are basically where *most* of the chemical action takes place, so it's important to have a fundamental understanding of what they are. Examples of the essential biological functional groups are listed in the next figure.

Hydroxyl (-OH)	Amino ($-NH_3$)
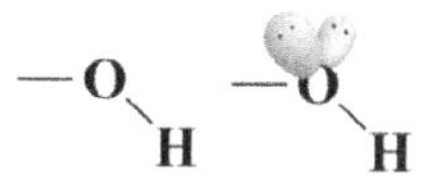	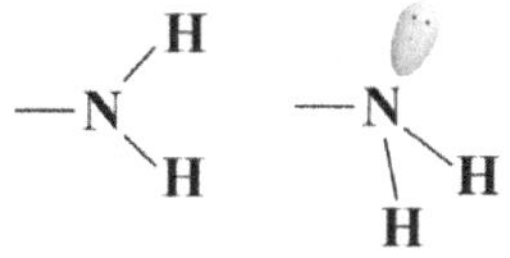
Carboxyl (-COOH)	Phosphate ($-PO_4^{2-}$)
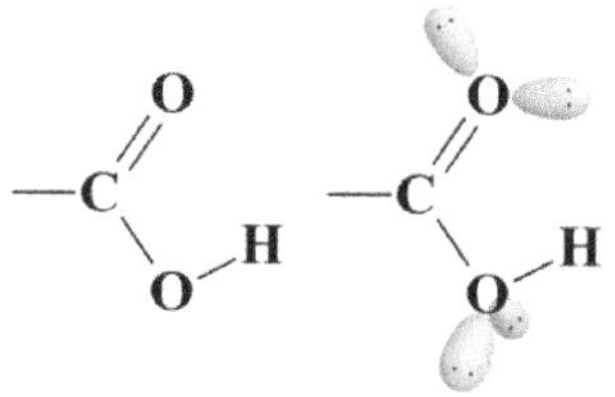	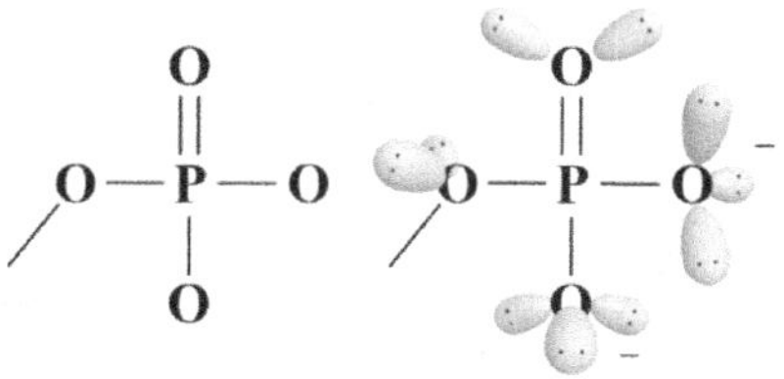
Methane ($-CH_3$)	Sulfide (-SH)
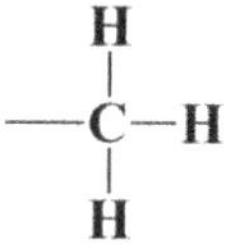	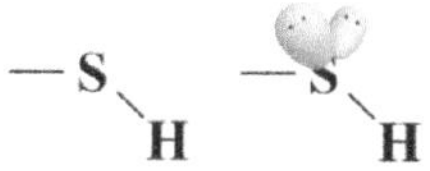

That's all well and good, but how do these molecules get made inside living organisms such as you and me? Do we literally take individual carbon, hydrogen, oxygen and nitrogen atoms from the environment and put molecules together one atom at a time? No. I can't remember the last time I found an individual carbon atom sitting all by itself waiting for someone to pick it up. For all practical purposes, carbon is found in the environment *already* bound into all manner of molecules. So, if a living system needs carbon, it will have to process away all the other atoms that are tagging along. That's not to say that we can eat just *any* carbon-based molecule, strip away the junk and get the carbon from it. Sometimes a molecule can have a structure that makes it poisonous. So, we stick to the things our bodies know how to process: **sugars**, **fats**, **proteins**, etc. We have the chemical machinery to break these down.

And break them down we do! The food we take in gets split into smaller and smaller pieces, releasing both the atoms and the energy locked in the bonds, until eventually all the carbon that was there gets turned into CO_2, and the hydrogen and oxygen gets turned into H_2O. The nitrogen gets packaged into ammonia (NH_3). This is the process of **metabolism**, which we'll talk about in later Topic Texts. Here's an overview image to wet your whistle.

Often, big complex molecules are broken down into a discreet set of specific molecules, such as **glucose**, **amino acids**, **fatty acids**, and so on. These smaller molecules, which we are very adept at using, get strung together one at a time in an organized fashion to make bigger molecules. It's like the bricks of a house; you put them together to make a grander thing. In the jargon of stuffy scientists, these molecular 'bricks' are called **monomers** and the bigger molecules they make are called **polymers**. In other words, polymers are just long chains of monomers put together.

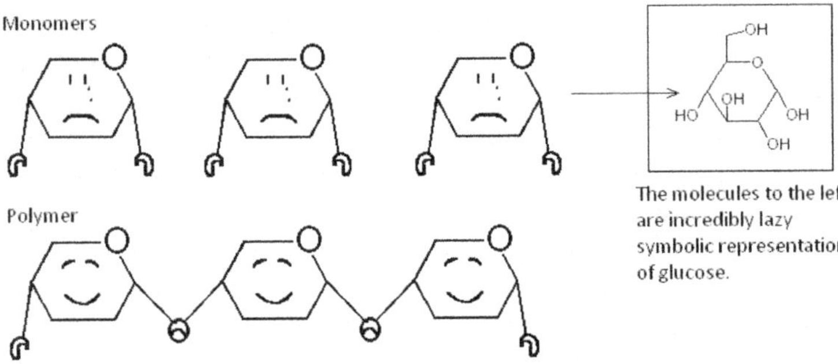

The molecules to the left are incredibly lazy symbolic representations of glucose.

The single most important type of reaction for stringing monomers together is called **dehydration synthesis**. You are probably familiar with the idea of dehydration from the common experience of exercise. It means, in a manner of speaking, "removing water," only here it does not occur because organic molecules go for a 5K run. Dehydration synthesis occurs when two molecules that contain hydroxyl groups, (or one hydroxyl and one amino group), are paired together and water is ejected, leaving the two parent molecules joined where the hydroxyl groups used to be. Here is a picture to help illustrate what happens:

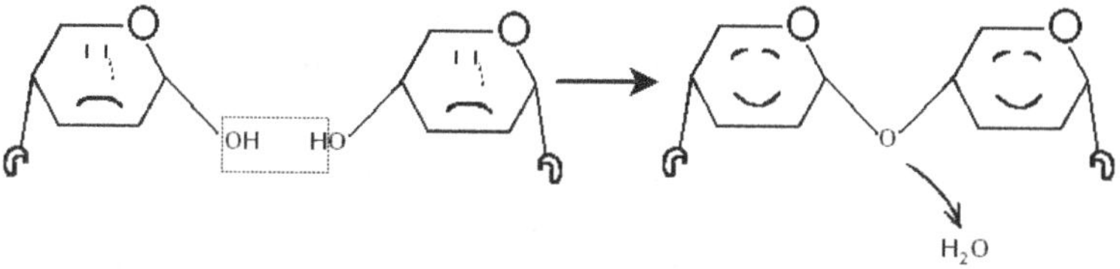

This reaction is the basis for almost all biological polymer synthesis, and it is well worth knowing. The way these polymers are broken down is in the opposite fashion, a reaction called **hydrolysis**, which means roughly "to break with water." In that case, it looks basically the same, only going in the reverse direction.

So, now that we know this little bit of organic chemistry, we are set to talk about the four main biological molecule types: **carbohydrates**, **lipids**, **proteins** and **nucleic acids**. That will be the focus of the next several sections.

Summary

In this section, we covered the basics of organic chemistry.

- Organic molecules are carbon-based molecules.
- Carbon is the backbone of biological molecules, since it can connect up to four atoms at a time.
- When chemists draw organic molecules, they represent the carbons as the corners in a stick model and leave out the hydrogen atoms. The number of hydrogen atoms present is four minus the number of lines attached to the corner.
- Functional groups are specific collections of atoms that typically appear attached to organic molecules. They are usually where the chemistry takes place.
- Common functional groups in biology are -OH (hydroxyl), -NH_2 (amino), -COOH (carboxyl), -PO_4^{2-} (phosphate), -CH_3 (methyl) and -SH (sulfide).
- Carbon, oxygen, hydrogen, nitrogen and other important elements come already attached to other atoms in the environment. In order for living systems to use them, they have to be able to deal with the other stuff that is attached. Unfortunately, no living system can process all the possible molecules that are out there, and many molecules can be poisonous.
- Living systems extract energy from organic molecules using metabolism. CO_2, H_2O and NH_3 are typical products of that process.
- Certain molecules found as monomers (or made into monomers from other molecules) are strung together to make chains of monomers called polymers. This process occurs through dehydration synthesis, in which water is removed each time two monomers are connected to each other.
- Hydrolysis is basically the reverse of dehydration synthesis, in which water is added to break two monomers apart from each other.

Unit 8

Carbohydrates

In this section we cover:

1) Monosaccharides: construction and uses
2) Basics of monosaccharide isomers
3) Polysaccharides: construction and uses
4) The water loving nature of sugars

Monosaccharides

Carbohydrates (a word that means "carbon and water") are biological molecules that are comprised of repeats of carbon, hydrogen and oxygen in a ratio of 1:2:1, or in other words, multiples of H-C-OH. The molecules that result are called **monosaccharides**, meaning "single sugars" in Greek. Monosaccharides typically have three to six H-C-OH repeats in them. Most monosaccharides are used in simple energy reactions, while the larger ones, (those with five or six H-C-OH repeats), can also be strung together into large polymers to serve the roles of energy storage, structural supports, or even information storage (such as in DNA). Of the monosaccharides, the 6-carbon molecules are the most numerous and important, so we'll focus on those during this unit. Before we do, however, let's have a brief look at a few of the smaller molecules to get our feet wet on the topic.

The image below shows three molecules, each produced by H-C-OH repeats. It may not look like repeats of H-C-OH since the

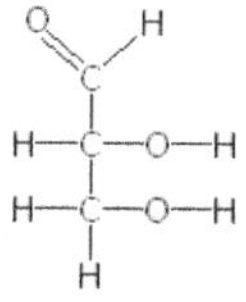

Glyceraldehyde

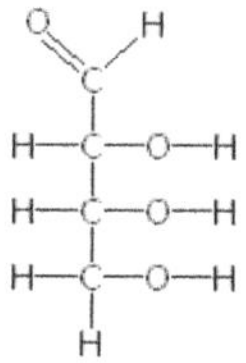

Erythrose

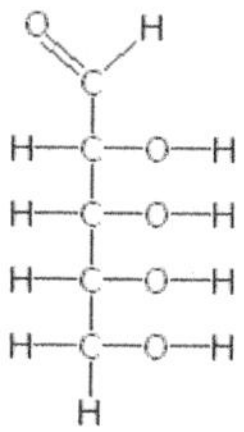

Ribose

carbons at the top are double-bonded to oxygen, but if you count the carbon, hydrogen and oxygen atoms, the ratio is indeed 1:2:1. Glyceraldehyde is an intermediate in energy metabolism. You'll see it again in future biology studies. Erythrose has a number of minor roles in living systems, but they are outside the scope of this book. Ribose is used to make **DNA**. We'll see it again in a few units.

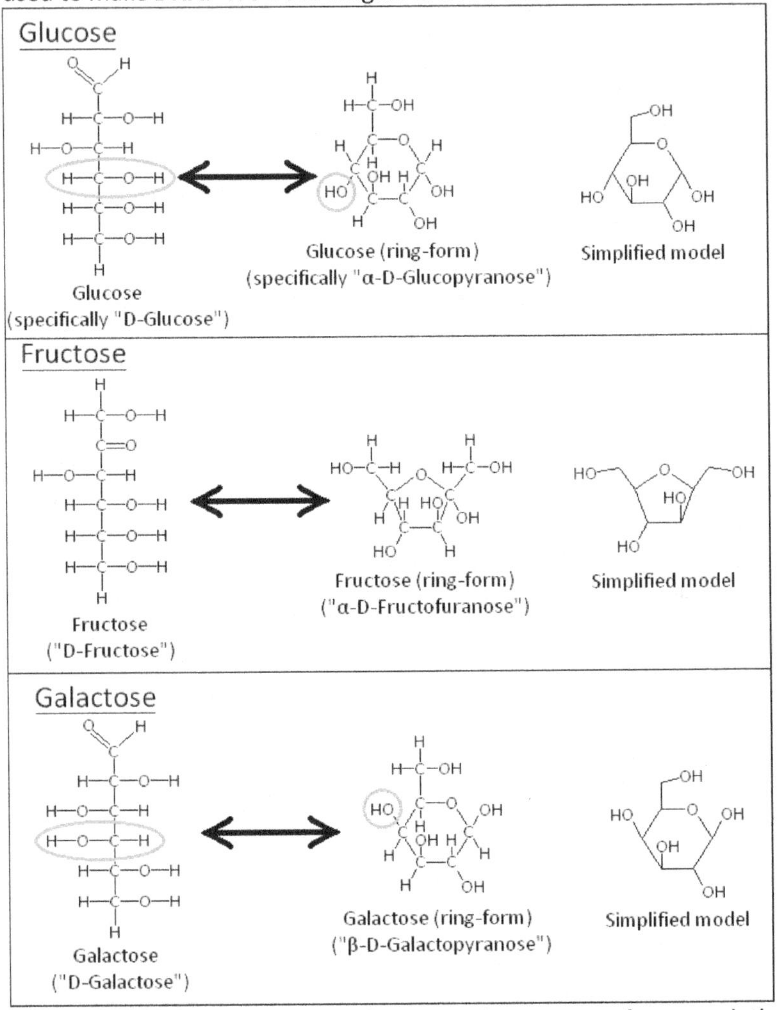

Having touched upon the general structure of sugars, let's zoom in on a few of the specifics of the 6-carbon monosaccharides.

These molecules can be ring-shaped or straight, switching back and forth spontaneously in living systems. Most of the time, they will be in the ring-shape, so I will usually draw them in that form for the remainder of this book. Examples of some common 6 carbon monosaccharides are provided in the image on the previous page.

Notice in that figure that simple sugars can differ from one another in the orientation of the side groups. For example, when you switch the direction of the third H-C-OH in glucose from right to left, it changes the molecule to galactose. It may not seem like it should matter, but this subtle switch gives the molecule completely different chemistry inside the living system. Molecules that have the same chemical formula but different orientation of side groups are called **isomers**. So, glucose and galactose are isomers of each other. There are many other possible isomers within the group of 6-carbon monosaccharides,[10] but we won't delve too far into the topic except to say that some types of isomers can be divided into two groups: "D" and "L."[11] The D's are mirror-images of the L's. For example, what we have been calling "glucose" is really "D-glucose," and its mirror image is "L-glucose."

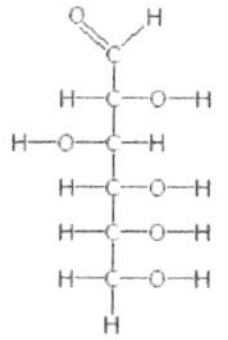

D-Glucose. This is the glucose we eat.

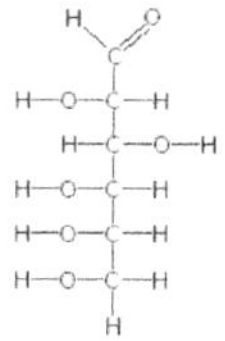

L-Glucose. This cannot be metabolized. High levels can be toxic.

---

[10] It's a bit sketchy as to how many isomers actually exist for six-carbon monosaccharides. Theoretically, it should be 32. In reality, the number of naturally occurring isomers is probably somewhere between twelve and 24. Once in a while, a strange bacteria, plant, or critter will be found to use an unusual monosaccharide isomer, but by and large, most organisms seem to use the same basic set of twelve.

[11] Most of the natural monosaccharide isomers are in the D group. D is short for "dextro" and L is short for "levo." Dextro monosaccharides twist polarized light to the right, while levo monosaccharides twist polarized light to the left. If this sounds Greek to you, don't sweat it. It's way outside the scope of general biology. You learn more about polarized light in courses on physics.

The two molecules certainly look like they could be the same thing, but they aren't! If you were to make a physical model of these two chemicals (an exercise you would do in a course like organic chemistry) you would see that it is impossible to superimpose them. They are definitely not the same. D-Galactose has a mirror image too: L-galactose. A picture of it would be displayed in a similar manner with a little mirror-like reflection showing a molecule that has oppositely positioned side groups.

So far, the six-carbon molecules that we've looked at are only a few isomers out of the total. We've barely scraped the surface! It is very easy to get bogged down by the details and lose our way, though. The main point to understand is that there are many ways to put together H-C-OH repeats, leading to many different possible molecular structures. Each one is chemically unique. The more of the H-C-OH repeats you put together, the greater the number of possible isomers that can be made. The existence of these isomers makes the carbohydrate class of biomolecules a very detailed class. Most basic biology courses do not go in depth in the discussion of monosaccharide isomers, however, as this topic is more properly pursued in courses on organic chemistry and biochemistry. If these courses are in your future, you can put off the torture until that time. If you do not intend to take those courses, you can put off the torture indefinitely. Either way, it is not necessary to know all the isomers in order to understand general biology. Presently, just be aware that the isomers are there, and it creates a *large* number of possible sugar molecules for living systems to use. That said, we're now going to see how these six-carbon sugars get strung together to make polymers called **polysaccharides**.

Disaccharides

The smallest polysaccharides that can be made from simple sugars are the **disaccharides**, which are just two monosaccharides linked together. These are made with dehydration synthesis. Some examples follow:

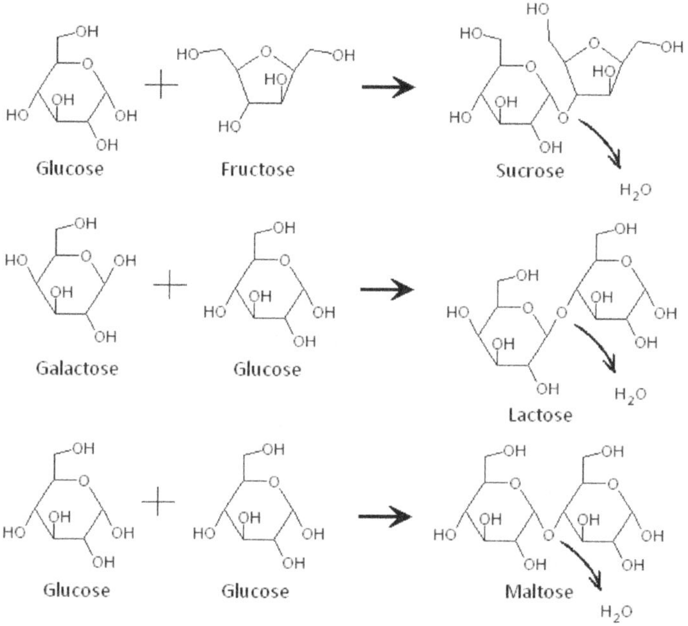

Like many monosaccharides, most disaccharides are also used as energy sources, and many plants use them for longer-term energy storage. Lactose is found in breast milk and is an important source of energy for young developing mammals. Sucrose is the common table sugar. Maltose is obtained from grain, such as barley, and is used to make malt beverages. Animals break these disaccharides down into their individual monosaccharides by hydrolysis at the -O- bond between the two monomers.

Oligosaccharides

As we string more monosaccharides together, we arrive at polysaccharides known as **oligosaccharides**, or "many sugars." These have a variety of uses in Nature, from energy storage to forming the basic structure of many antibiotics, to being the source of alcohol produced in some beverages.[12]

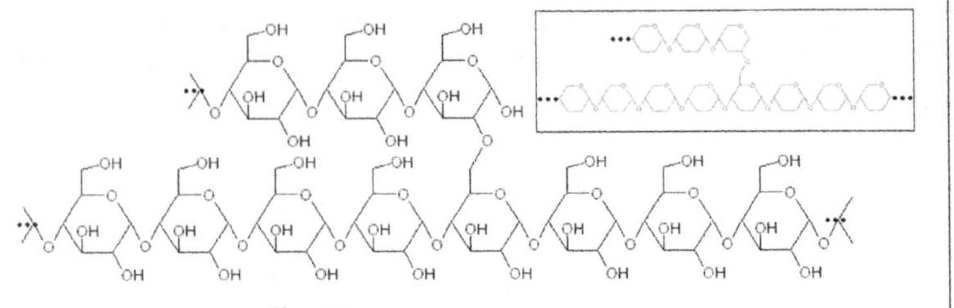

Glycogen

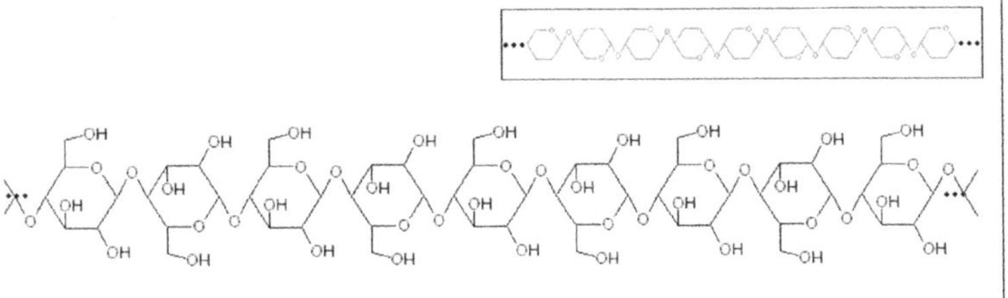

Cellulose

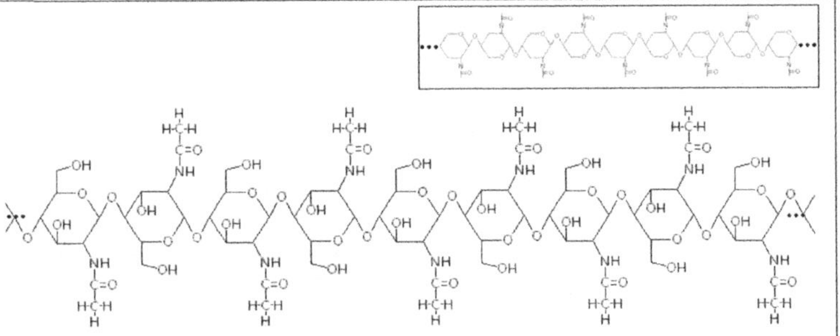

Chitin

Long Chain Polysaccharides

Long chain polysaccharides have two general uses in living systems. Some polysaccharides serve as long-term fuel sources (i.e., starch in plants and glycogen in animals). Other polysaccharides, such as cellulose and chitin, serve as structural molecules. Cellulose is made in plants and gives their stems rigid strength. Chitin, a

[12] The main component of alcoholic beverages is ethanol, CH_3CH_2OH, known commonly as "alcohol." It is not a monosaccharide, but is made from monosaccharides by microscopic organisms.

polysaccharide that also has nitrogen groups in it, is made by insects and crabs to form their tough outer skeletons (**exoskeletons**). The previous image shows what glycogen,[13] cellulose and chitin look like.

One should remember that although the basic unit of polysaccharides is the "simple sugar," this does not imply that all polysaccharides can be eaten and digested. The reality is quite the opposite. Apart from starch and glycogen, very few polysaccharides are actually digestible. In fact, only a handful of organisms (mostly microscopic ones) are capable of breaking down polysaccharides such as cellulose and chitin. Though you can eat lettuce, the cellulose in it does not break down into its respective sugar pieces, but instead gets passed through the gut. You can be thankful for that, because if we COULD digest cellulose, the actual number of calories in a salad would be horrendous.

The Water Loving Nature of Carbohydrates

If you take another look at any of the molecules we studied in this section, you'll notice that they all have a bunch of -O- and -OH groups. Each oxygen has lone pairs sticking from its side, (although they were not drawn explicitly). In addition, the -OH groups are notorious for participating in hydrogen bonding. You may remember that in an earlier unit we noted that water also engages in a lot of hydrogen bonding. Taken together, these qualities make carbohydrates attract a LOT of water, so much so that large polysaccharides are like molecular sponges. Long chains of cellulose, for example, are typically laid down next to each other and interconnected, and when the water infiltrates these water-loving strands, the entire structure swells and becomes incredibly rigid. It is this behavior that helps give cellulose its structural abilities. Glycogen attracts a lot of water as well, but it is not cross-linked in the same manner. It's connected together in the middle and allowed to spread out, branching here and there. The result is that it swells up like a big molecular ball, with each individual polysaccharide strand sticking out into the surrounding water. Since there are a large number of strands sticking outward, there are a lot of possible sugar monomers that can be broken off and used for energy at any given point on the ball's

[13] Starch looks very similar to glycogen.

surface. While this means that energy can be released from glycogen very quickly, it also means that glycogen takes up a LOT of room. If a living system needs to store a LOT of energy, it would quickly run out of room if it chose to store all of it as glycogen. So how does a living system store a lot of long-term energy? The answer to this dilemma lies in a different class of molecules called lipids, which we will explore in the next section.

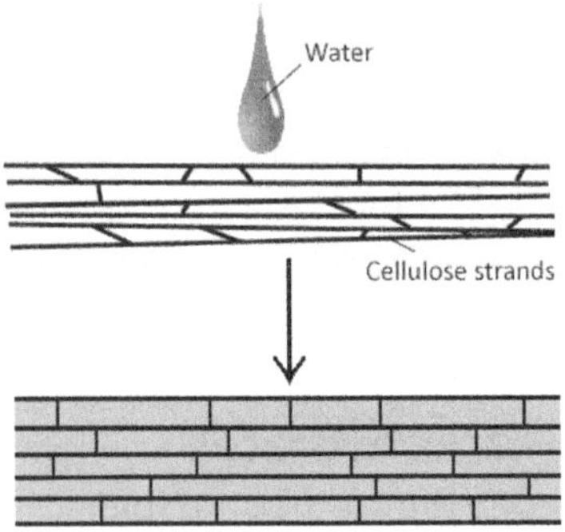

Cellulose with water throughout. Water floods between the strands and stengthens the entire structure. If only it worked for me too.

a) A glycogen molecule. Notice how the individual strands grow from a central point (which contains a protein called glycogenin). Also notice how the individual branches split here and there.
b) Also my hairstyle in highschool. Ah, the good ol' days.

Summary

To summarize this chapter:

- Monosaccharides are a class of molecules made by putting together repeats of H-C-OH. Their overall structure has a carbon to hydrogen to oxygen ratio of 1:2:1.
- Monosaccharides are largely used in energy management. Most of the small monosaccharides are used primarily for this, though they have some other minor uses as well.
- Monosaccharides have a lot a different ways they can be put together, creating isomers of each other. Monosaccharides are broadly divided into "D" and "L" isomers. D and L isomers are mirror images of each other. Living systems use mostly D isomers.
- Six-carbon monosaccharides have a lot of isomers, which are also broadly divided into the naturally occurring "D" isomers and the rare "L" isomers.
- The six-carbon monosaccharides are often linked together by dehydration synthesis to make polymers called polysaccharides.
- Disaccharides are polysaccharides with only two monosaccharides linked together. They are used primarily in energy processes.
- Oligosaccharides are a little bigger than disaccharides and are used primarily for energy. They also have smaller roles as the structural basis for some antibiotics, as well as a few other miscellaneous uses.
- Long chain polysaccharides are used for long-term energy storage, but some have structural roles.
- Glycogen is an important energy-storing polysaccharide found in animals.
- Starch is an important energy-storing polysaccharide found in plants.
- Cellulose is a structural polysaccharide made by plants. It gives them their rigid stems and leaves.
- Chitin is a nitrogen-containing polysaccharide derivative that is found in bug and crab shells. It gives the shells their hardness.
- Carbohydrates attract water. This is why cellulose networks swell and get rigid, and why blobs of glycogen spread out and take up lots of space.

Unit 9

Lipids

In this section we cover:

1) Fats, oils and waxes: their uses and characteristics
2) Phospholipids and how they form layers in water
3) Steroids: structure and uses

Lipid Subcategories

Lipids are mostly non-polar molecules that are made up of primarily carbon and hydrogen, and they are constructed from repeats of $-CH_2-$ linked together. There are three broad categories of lipids: 1) fats, oils and waxes, 2) phospholipids and 3) steroids. Each category tends to have specific uses.

Fats, Oils and Waxes

Fats, oils and waxes are made up of two general parts: 1) a **glycerol** molecule, and 2) one or more **fatty acid** chains. A fatty acid chain is a long molecule that has numerous $-CH_2-$ repeating units with a carboxyl group on one end. These are combined with the glycerol molecule through dehydration synthesis. Three such fatty acids and a glycerol make a **triglyceride** molecule, as shown to the right:

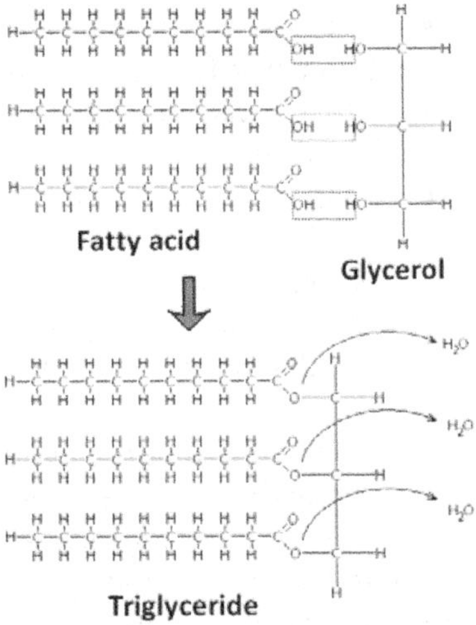

When we discussed carbohydrates, we talked about how storing energy as glycogen has the unfortunate drawback of taking up lots of room. The sugars in polysaccharides like to stretch out into the water, soaking up the water molecules like a sponge. Lipids do the opposite. They hate water and shrink away from it. Also, the C-H bonds in lipids are very energetic (meaning they can store a lot of energy to be used later by the living organism). With these two qualities together, lipids serve as a very efficient form of biological energy storage. Triglycerides like that shown on the previous page are a very common molecule used to store energy.

Let's look a little closer at fatty acids. In the first image, we saw little fatty acid chains that were ten carbons long, but fatty acids can have variable numbers of carbons in them. The majority have an even number, however, (i.e. 10, 12, 14, 16, and 18 "$-CH_2-$" repeats, etc.). Fatty acids that have double bonds in them are called **unsaturated**. Those that have only single bonds linking all the carbons together are called **saturated**. This is the origin of the terms "saturated fat" and "unsaturated fat."

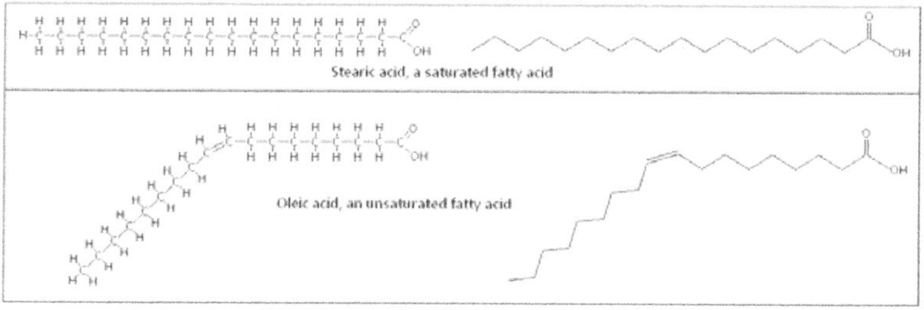

Stearic acid, a saturated fatty acid

Oleic acid, an unsaturated fatty acid

There are two consequences to fatty acids being saturated. First, their non-polar chains can straighten out and nestle against each other, packing tightly. This not only means more efficient energy storage, but it also means they tend to be more solid or semi-solid at room temperature, like lard or shortening. Second, saturated fatty acids have more energy than unsaturated fatty acids, because C=C double bonds are not as energetic as the C-C single bonds.

Fats (such as the kind stored in our rumps) have mostly saturated medium-length fatty acid chains (16, 18, or 20 carbons, etc.). Oils, on the other hand, are shorter, mostly *un*saturated fatty acids or are short hydrocarbon chains (linear molecules with nothing but carbon and hydrogen). Since they are very small and many do not pack very efficiently with each other, they are liquid at room temperature. Oils have a wide variety of uses in biology, from energy storage to surface waterproofing to chemical defense. For example, the oil on skin traps water inside the body, helping to keep us from drying out. The crude oil that is found deep in the ground is a collection of all the biological lipids that have deposited over the billions of years of life on the planet. All that fat from poor T-Rex's love-handles has become buried, and we use the unspent biological energy to run our vehicles. T-Rex's lizard butt now helps to haul our big butts around.

A bunch of saturated fatty acids packing very closely to each other. One unsaturated fatty acid stands out in the middle, unable to pack as closely because of the double-bonded kink in its tail.

Waxes tend to have *very* long fatty acid chains. This makes them stick to (and tangle) together with each other more tightly than fats and oils. Consequently, they form solids at room temperature. Waxes are not edible by most animals but serve instead as waterproofing and structural support in things like plants, animal furs and beehives. Earwax, for example, provides a basic defense against bacteria, fungi and insects. It also does an incredibly good job at annoying us on a regular basis. Precisely how cavemen managed without cotton swabs is beyond me.

Phospholipids

Phospholipids are very similar to trigylcerides in that they have a glycerol backbone. Instead of three fatty acid subunits, however, they have only one or two. Like triglycerides, these subunits are connected by dehydration synthesis. In place of the third fatty acid, it has a very polar side group that often contains phosphate and nitrogen. They come in many forms, but here is a typical example:

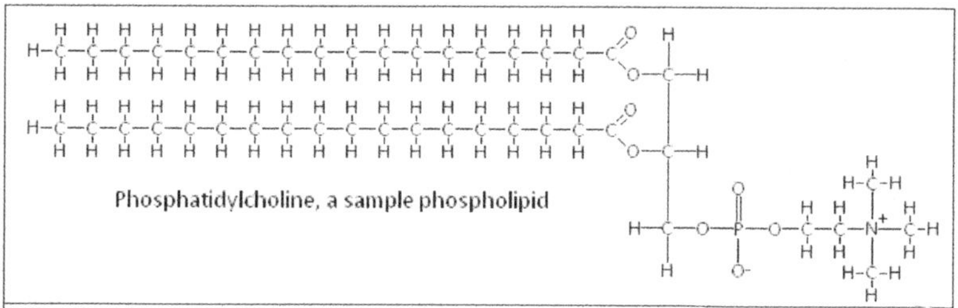

Phosphatidylcholine, a sample phospholipid

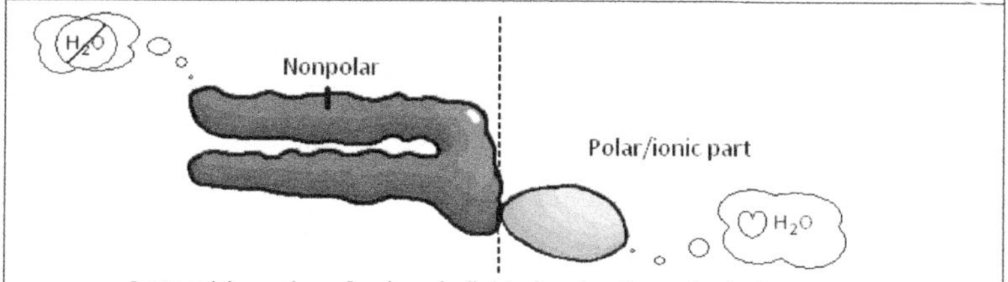

Cartoonish version of a phospholipid, showing the water-hating nonpolar tails and the water-loving polar head.

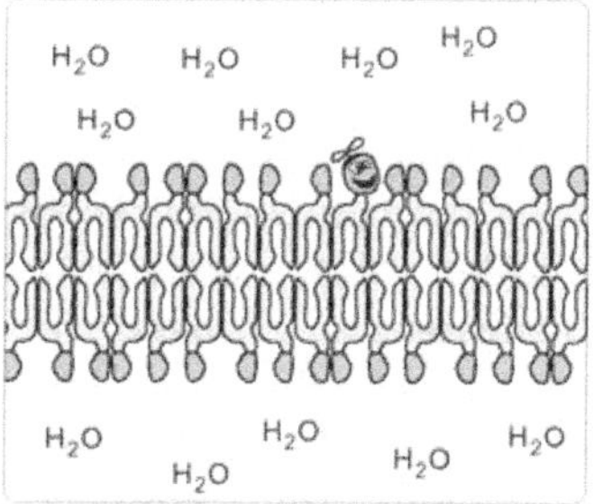

This unique structure has hydrocarbon tails that are hydrophobic and head groups that are polar (sometimes ionic) and thus hydrophilic. This causes phospholipids to line up side by side, with their polar heads pointing into the water and their non-polar tails tucked tightly against each other away from the water. This is a crucial feature for the development of the cell membrane, which is discussed in the next Topic Text.

Steroids

Steroids are unlike the other two categories of lipids in that they are not made from glycerol backbone and fatty acids. Rather, they consist of series of ($-CH_2-$)-containing rings based on the general design for cholesterol (shown in the next graphic).

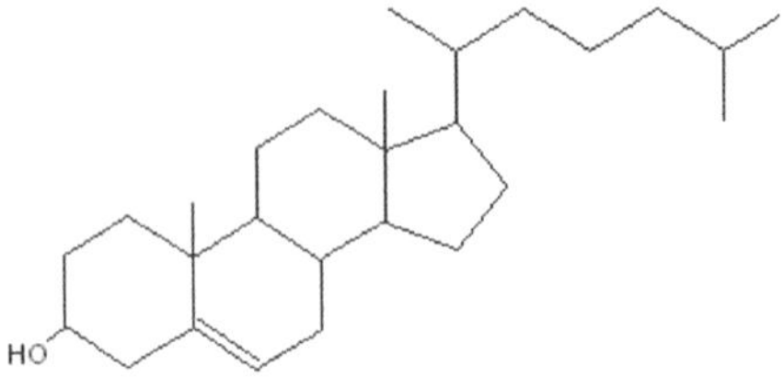

Cholesterol

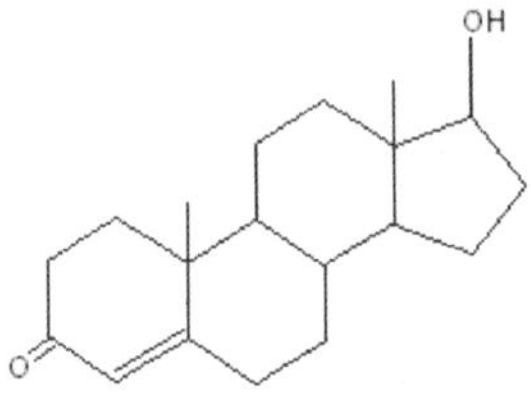

Testosterone

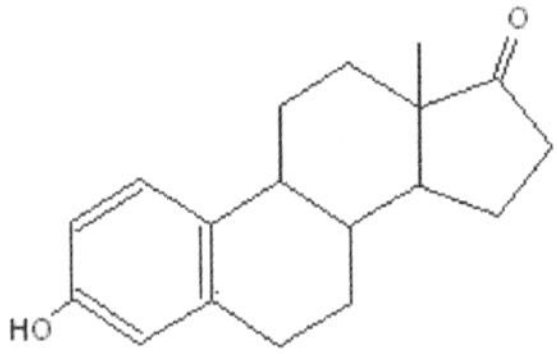

Estrogen

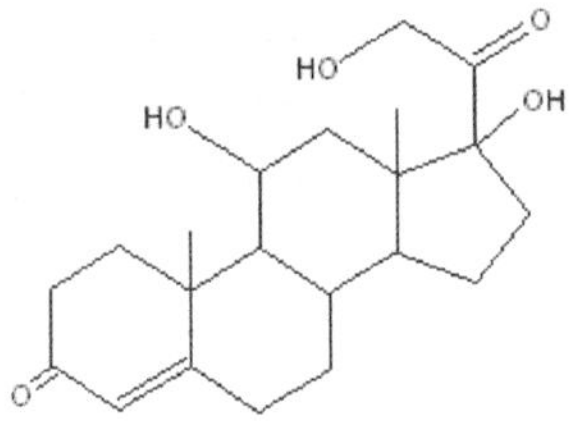

Cortisol

Steroids have a very diverse role in biology. For example, cholesterol can be found in cell membranes where it helps to keep phospholipid hydrophobic tails from packing too tightly. This keeps the membrane from becoming too solid. Other steroids are used as chemical messengers called hormones (**testosterone**, **estrogen**, and **cortisol** are examples), which signal distant parts of the body about changes in body conditions. Steroids, like cholesterol, can also serve as the starting point for the creation of several important molecules. These topics are generally the focus of later courses, however, such as second semester general biology, anatomy and physiology or biochemistry. We will not discuss them much further in this book.

Summary

In this section, we covered the basics of lipids.

- Lipids are mostly hydrophobic compounds divided into three main categories:

  1) fats, oils and waxes

     - Fats are semi-solid compounds used primarily for compact biological energy storage.
     - Oils are short fatty acid (or hydrocarbon) chains that are liquids at room temperature. They typically serve as energy storage and waterproofing.
     - Waxes are very long fatty acid chains that pack so closely to each other that they form solids at room temperature. They are used to protect/waterproof surfaces and to create structures such as beehives.

  2) phospholipids

     - Two fatty acid chains + 1 glycerol molecule + 1 hydrophilic head. These line up with each other so that their hydrophobic tails pack together, while their hydrophilic heads interact with water. This creates a double layer of molecules within the water and becomes the important basis for cell membranes.

  3) steroids

     - These are multi-ring chemicals based off of the cholesterol molecule. They have a diverse number of uses in biology, such as helping to keep the cell membrane more fluid, or serving as chemical messengers called hormones. Cholesterol is the starting point for the creation of several other important molecules.

Unit 10

Proteins

In this section we cover:
1) Protein roles
2) The 20 common amino acids
3) Protein structure: primary, secondary, tertiary and quaternary
4) Sulfur-sulfur bonds and proline bends
5) Protein domains
6) How protein structure relates to function
7) Structure denaturing
8) The number of proteins that can be made, and where this information is stored.

The topic of proteins is a very complex one. Here we discuss the basic concepts of the structure and uses. We will encounter proteins in a number of places throughout other Topic Texts and will pick up more detail as those topics arise.

Protein Roles

Proteins are biological polymers that have very important roles in living systems. First and foremost, they can act as **enzymes**. Enzymes are catalysts - they speed the rate of other reactions. Enzymes are specific and selective in this task, each particular enzyme speeding only one particular reaction or a set of similar reactions. Second, proteins can serve as microscopic structural supports. Third, some proteins can be **contractile** (changing shape upon stimulation), serving as the means by which muscles shorten (we'll discuss the specifics in a later Topic Text). Fourth, they can serve to transport chemicals rapidly from one place to another inside living systems. Fifth, they can be used as sources of nutrition for the young, such as might be found in breast milk. Sixth, some proteins are used as

hormones. Hormones are chemical messengers that link together physiological systems. We talked very briefly about this topic when we looked at the cholesterol-based steroids. Last, and certainly not least, proteins make up **antibodies** and most of the other components of the immune system, which fights off infections and helps to keep us healthy. Hormones and antibodies are usually covered in advanced biology classes.

Protein Roles - Overview
1) Enzymes – Speeds the rate of reaction of specific reactions.
2) Structural units – microscopic framework upon which other molecules can attach, or upon which reactions can take place.
3) Contractile units – the primary mechanism that acts to shorten muscle.
4) Transportation – efficiently carry materials through the body.
5) Nutrition in milk – helps to provide nutrients to offspring.
6) Some Hormones – chemical messenger that acts to inform physiological systems of external stimuli.
7) Antibodies – helps fight off infection.

Protein Synthesis, the Basics

Proteins are polymers. The monomer building blocks used to make these polymers are called **amino acids** (commonly abbreviated aa's). All amino acids are related by a common backbone chain of three atoms: one nitrogen and two carbons, as show here:

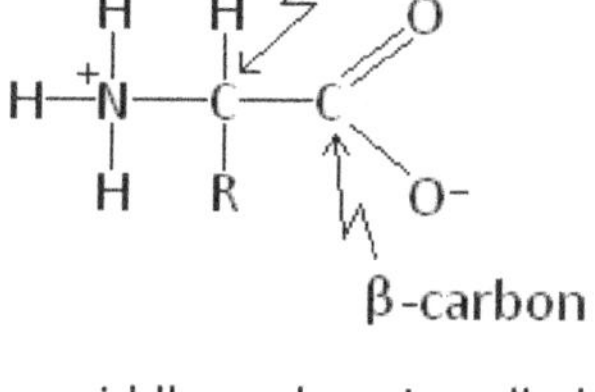

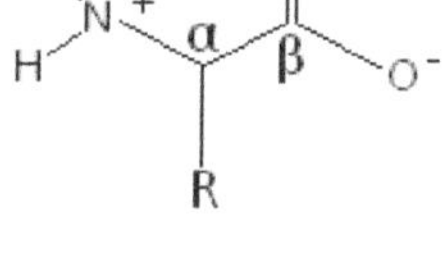

The middle carbon is called the **alpha-carbon**. The end carbon is called the **beta-carbon**. Attached to the alpha-carbon is a side group, labeled "R" in the picture. Each amino acid has a unique R group, and it is this R group that determines the unique chemical behavior of that amino acid. There are twenty fundamental **amino acids** that living systems have in common. In the image below you can see that the R groups range from very simple (just an H for glycine) to very complex, such as the rings in tryptophan and the long nitrogen chain in arginine.

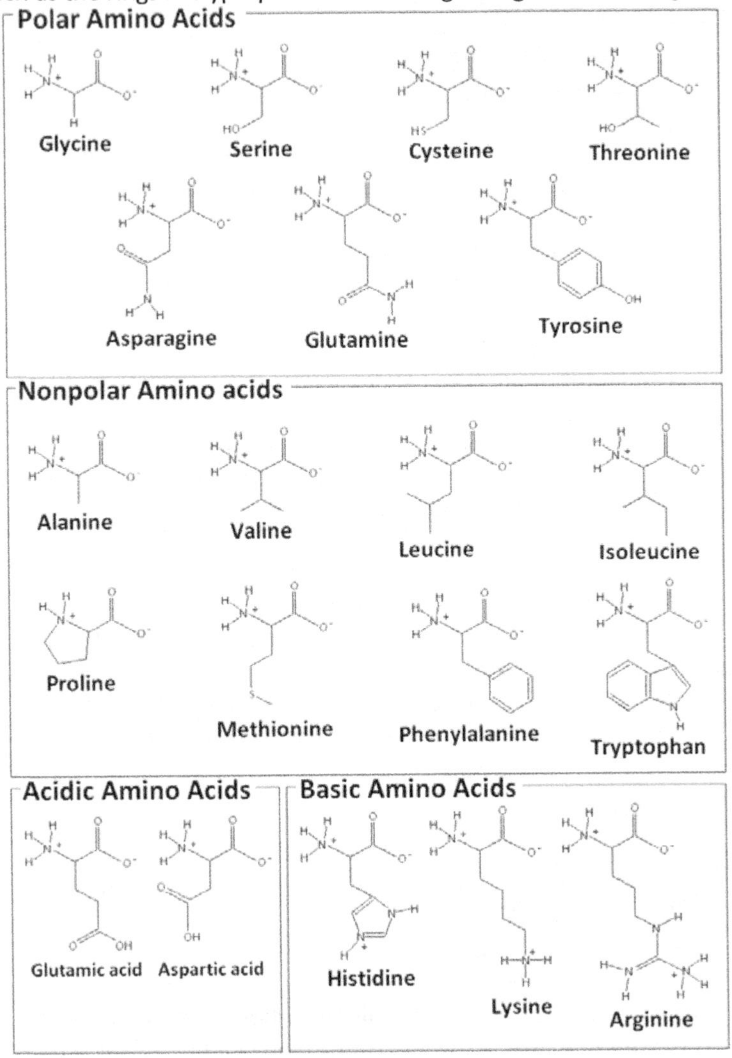

You'll notice in this image that each amino acid backbone has a positively charged nitrogen on the left and a negatively charged oxygen on the right. The reason for this is that the COOH functional group is acidic (tending to release its H^+ to become COO^-, as shown), while the NH_2 group is basic (tending to attract H^+ to become NH_3^+). So, individual amino acids are typically drawn with COO^- and NH_3^+.

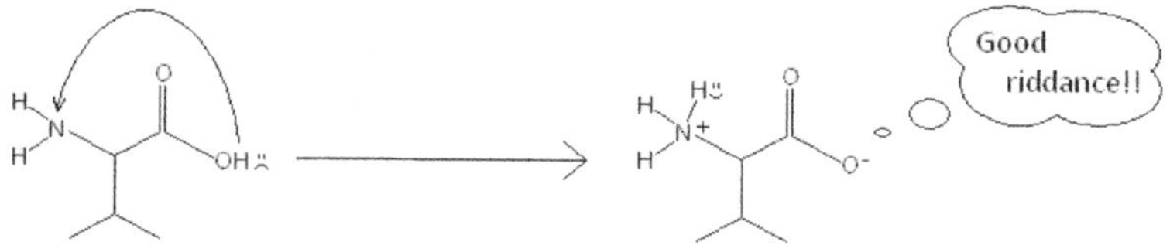

Valine Valine

Since the hydrogen that is released from the COO^- on the right is actually H^+, you might be wondering why I put the "+" next to the nitrogen in the drawing. This is because of the way the electrons are shared between the nitrogen and H^+; the positive charge tends to gather over the nitrogen.

The next thing to notice in the big image of all the amino acids is that they are divided into four broad categories: nonpolar, polar, acidic and basic. These categories refer to the characteristics of the amino acids when they are bound together into proteins. To shed light on that, let's look at how amino acids are connected together.

Like the previous polymers we studied, amino acids are strung together using dehydration synthesis. The easiest way to see this is to draw the amino acids the same as the valine in the last image – with the H still attached to the oxygen on the right. The reason is to avoid confusion, since the dehydration synthesis reaction is otherwise not very obvious. The real reaction turns out to be the same regardless of how the amino acids are drawn, but it involves some otherwise unimportant details that will only serve to confuse rather than to educate. At any rate, the result is the usual: water is released when the monomers are connected together. The following picture shows three amino acids getting connected into a small protein polymer:

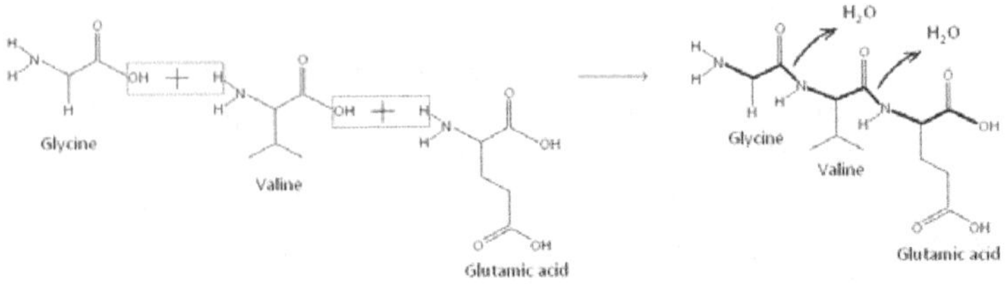

This little protein (also called a **peptide**)[14] is only three amino acids long, with a knobby, crooked backbone connecting each monomer (drawn as dark thick lines). We could easily extend this to any length we like, of course. For example, here is a cartoonish protein with thirteen generic amino acids, (i.e., side group "R"), drawn with the backbone as a simplified straight line.

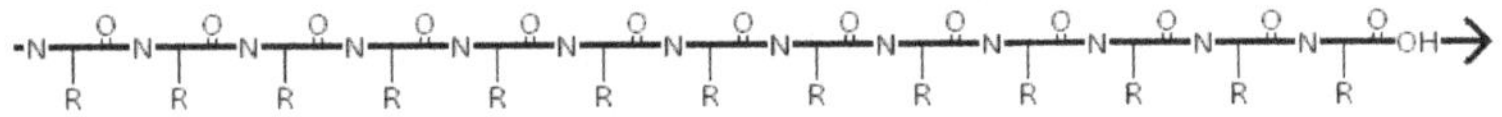

Peptide chains sometimes have a directional arrow drawn on one end to make it clear which direction is the end with the COOH. In this last picture, the arrow points to the right because the right end of the protein has the COOH and the left end has the nitrogen.

As we saw in the picture of the three amino acids, the N-C-C backbones of proteins are *not* straight, but very crooked, bending at each atom because of the weird bonding angles they make. The strange shape of the protein backbone isn't the only determining factor in the overall shape of the protein, though. As it turns out, the backbone is very flexible, allowing the protein chain to fold into all kinds of bizarre shapes. This, in turn, leads to some interesting structures. . .

---

14 Peptide = one or more amino acids stuck together by this mechanism.

Protein Structure

The order of amino acids in a protein is called its **primary structure**. The backbone of the primary structure, as we learned, is an alternating pattern of N-C-C-N-C-C. . . and so forth. The nitrogen-hydrogen bonds in this backbone are polar, with a partial positive charge positioned over the H. In addition, the oxygen atoms of the repeating C=O bond have two lone pairs poking out from them. This makes them partially negative. So, if we zoom in on our cartoonish protein from the last image, we would see a whole bunch of partial positive and negative charges everywhere.

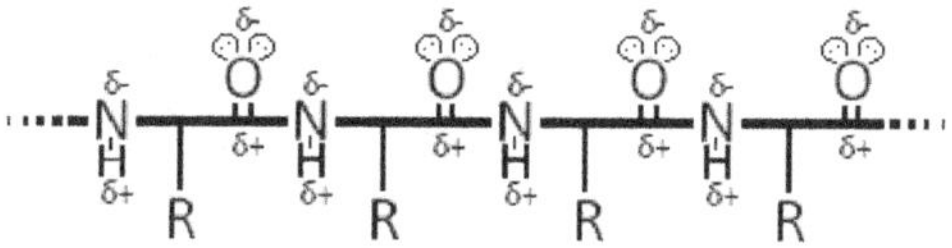

The result of all these polar bonds is that the protein chain is very sticky. It's like a shoe string that has gooey stuff all over it. Before too long it gets all knotted up, sticking to itself due to the partially positive sites being attracted to the partially negative sites. As it turns out, there are two common ways that protein chains bond in response to these attractive forces. The first is the **alpha-helix**. In this process, the chain starts to spiral around in such a way that a hydrogen from farther along the chain forms a **hydrogen bond** with an oxygen farther back, causing a corkscrew shape. Here is roughly what it looks like:

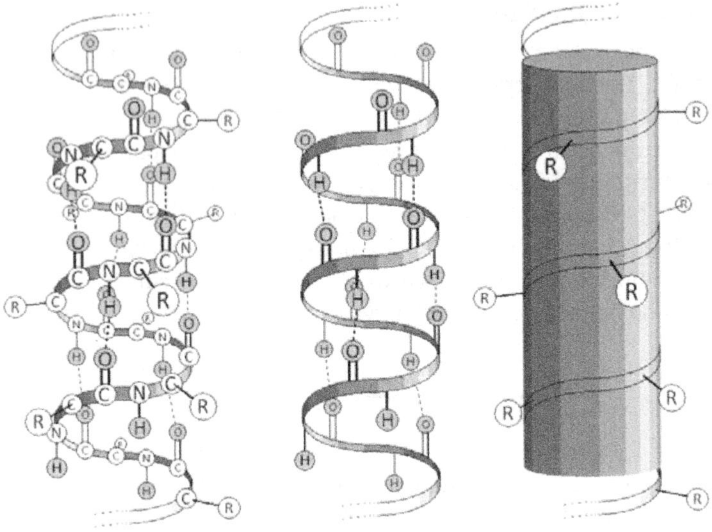

The far left shows all the atoms in their raw complexity. The ribbon that connects them is the backbone that runs along the chain. This, of course, is a cluttered picture, so have a look at the central image. This image shows the backbone ribbon with only the hydrogen bonding between the sticky hydrogen and oxygen shown as individual atoms. The cylinder on the far right is a stylized version of the alpha-helix; this is how it is typically drawn in many biology books. In that simplified version, you can see that the alpha-helix is basically like a tube where the side groups stick outward.

The second way that protein chains often organize is through the **pleated-sheet** or **beta-sheet**. In this process, the chain bends backwards and overlaps itself, like a person touching their toes.

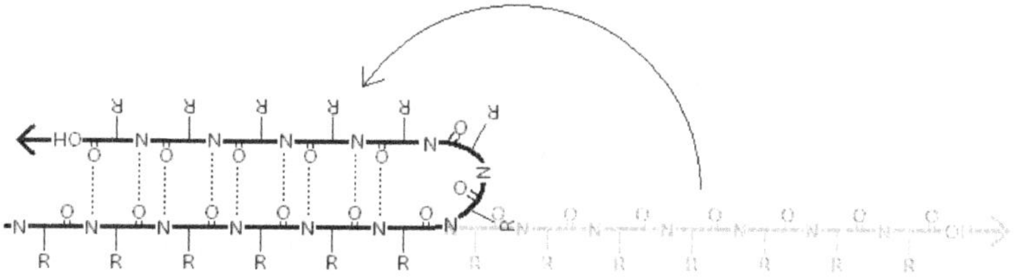

It looks fairly simple in this picture, but the actual 3-D structure is like a ruffled potato chip, or a wavy roof, with the "side" group sticking up and down.

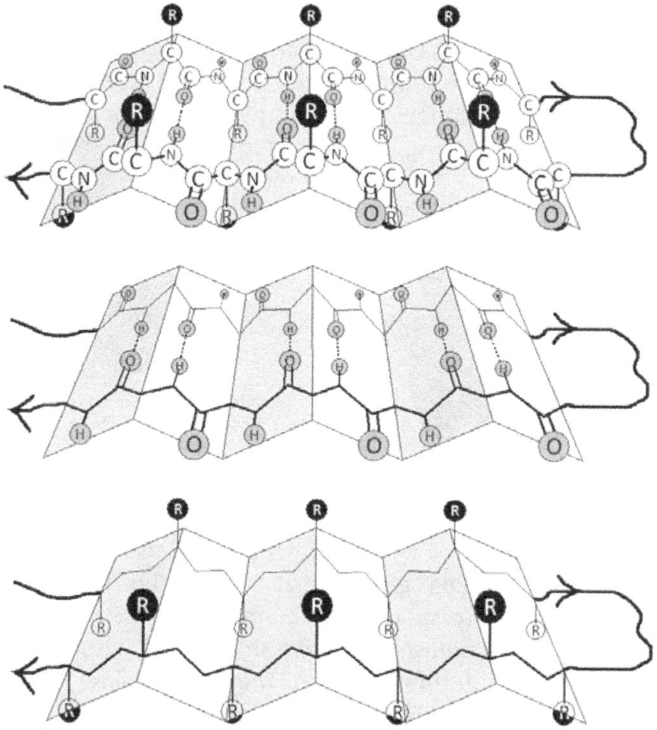

The bottom version of the beta sheet is typically drawn with flat arrow symbols, like this:

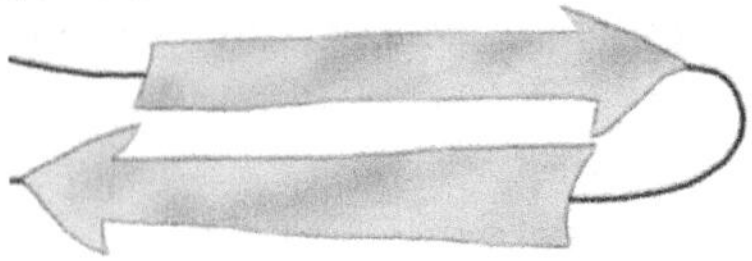

Whereas the order of the amino-acids is called the primary structure, the alpha-helix and beta-sheet structures are referred to as the **secondary structure** of the protein. These structures, in turn, tend to rest against each other to form the **tertiary structure** of the protein. The precise method in which all of this occurs is mainly dependent on the side groups that are sticking out (the R's). Amino acids that have non-polar side groups are hydrophobic and will cause the entire thing to arrange so that they are tucked inside the protein and away from water. Amino acids that have acidic, basic, or polar side groups are hydrophilic and will cause the structures to arrange so that they are pointing out into the surrounding aqueous (water) environment.

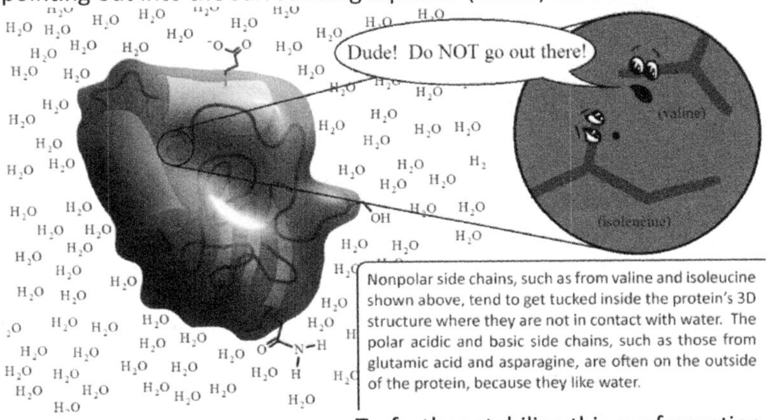

Nonpolar side chains, such as from valine and isoleucine shown above, tend to get tucked inside the protein's 3D structure where they are not in contact with water. The polar acidic and basic side chains, such as those from glutamic acid and asparagine, are often on the outside of the protein, because they like water.

To further stabilize this conformation, side groups that have sulfur in the them (cysteine and methionine) are clipped and adjusted so that a sulfur-sulfur (disulfide) bond forms between the two chains, like glue sticking a clump of hair together.

In addition to sulfur-sulfur bonds, we need to make note of the strange behavior of proline. It's shaped in a ring from the alpha-carbon to the nitrogen atom. This makes the backbone of proline bent. Proline, therefore, tends to bend the protein chain sharply wherever it appears. This makes it useful in positions where the protein backbone chain needs to bend at sharper angles in order to form a specific overall shape. To this end, proline is often found in sections that join helices and sheets together.

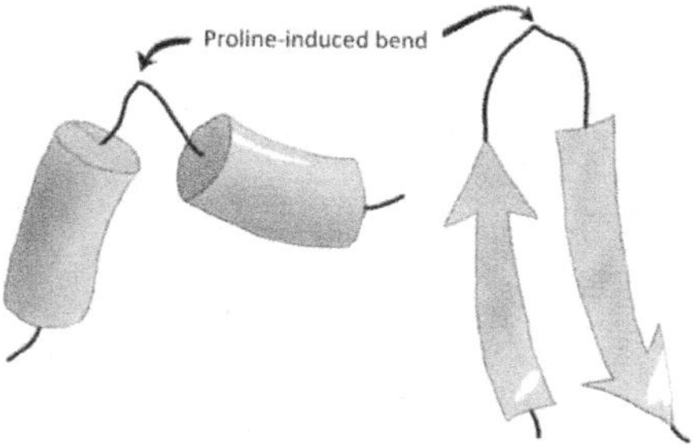

Together with the sulfur-sulfur bonds, the proline bends and the weird shape of the alpha-helices and beta-sheets, the protein chains look like blobs with tubes and flat arrows in them. Now, if multiple blobs clump together, (i.e. several individual protein chains), you arrive at the **quaternary structure** of proteins. Like the tertiary structure of proteins, the individual blobs are often held together by sulfur-sulfur bonds and connect in a way to minimize the number of nonpolar side groups that are exposed to water.

Here is an overview of protein structure:

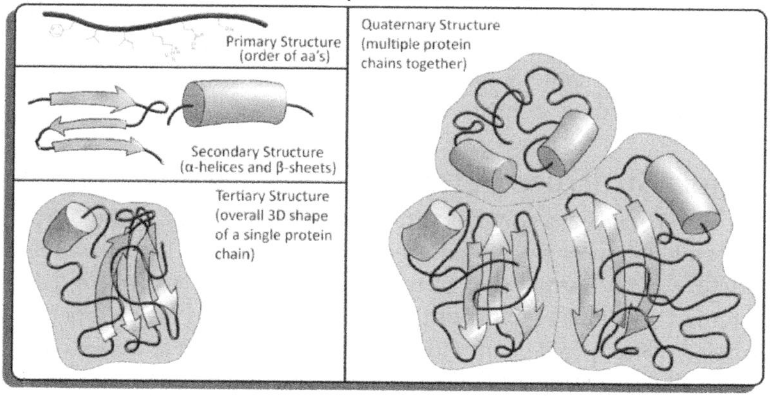

One final note on protein structure: domains. If you look at the overall quaternary structure of proteins, you will often notice discrete regions in the protein. We call these **domains**.

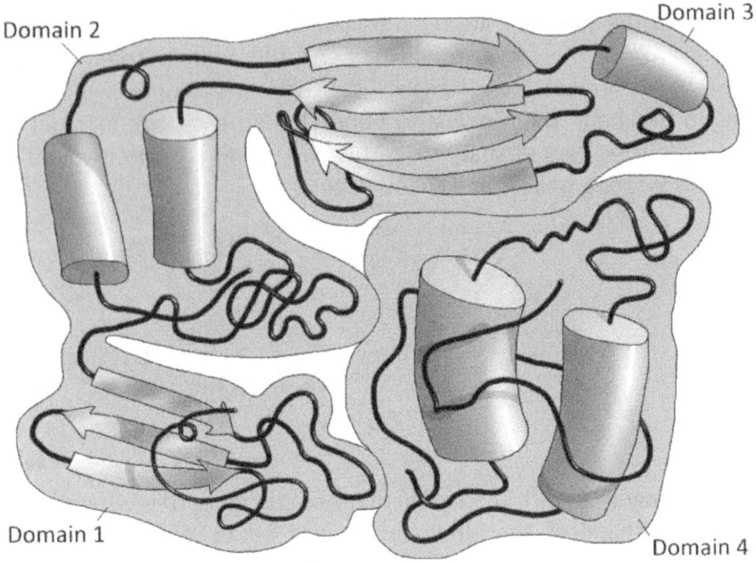

Domains can sometimes be made from different protein chains, but that isn't always the case. In this last image, three domains are made from one continuous protein chain. The fourth domain is made from a completely different protein chain.

Protein domains typically have their own chemical job they perform, and completely unrelated proteins can sometimes have very similar domains because parts of their chemical jobs are similar to one another. A full appreciation of the topic on protein domains requires understanding of genetics, evolution and biochemistry, so we'll put off further discussion of domains for this book.

Protein Structure Determines its Characteristics

Proteins are complex three dimensional things, created from the combination of twenty different building blocks. The overall 3-D structure of the protein, in turn, often determines its uses in living systems. Let's look at a few of the roles that proteins play. Structural proteins typically fold up into long and thin blobs, such as those in hair or in cellular scaffolding. Sometimes they connect together like a chain, extending long distances. The strands they form not only provide support for the living system but can also be used as a kind of road for other proteins to travel. The transport proteins, such as

dynein and **kinesin**, often look like little blobs with long legs and feet. These "leg" domains will attach to the long structural proteins and "walk" along it, carrying stuff for up to the entire length of the organism.

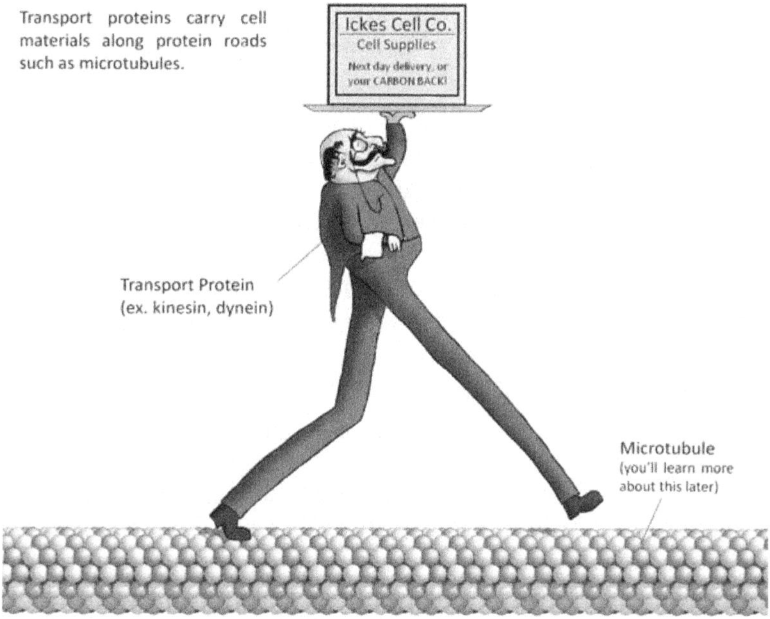

Transport proteins carry cell materials along protein roads such as microtubules.

Proteins that act as enzymes, on the other hand, are characterized as blobs that fold so that they have little nooks and pockets into which very specific molecules fit. By creating a very special environment for these molecules, it speeds the chances of reaction between the two molecules that fit into the space. Not just any molecule can fit into these little crevices, though. These spaces are very specific. How? Two general mechanisms make this possible: 1) The shape tends to be the same as the specific molecule, and 2) the charge (and/or partial charges) tend to be optimal to attract the specific molecule. As such, protein enzymes recognize their target molecules with expert precision and make sure that reactions between the target molecules happen much faster than they would otherwise occur.

Not all the roles of proteins are necessarily characterized by highly recognizable structure. Antibodies, for example, are a very diverse group, but the prototype structure looks like a capital "T", where the top ends of the T are shaped in a way that it specifically recognizes the surface of various bacteria, parasites and viruses. Exactly how this works will be discussed in a later book.

Protein Denaturation

The structure of the proteins determines its usefulness. The structure can be disrupted in a number of different ways, though. The first is to heat the protein. This causes the hydrogen bonds and the -S-S- bonds to snap, so that the protein becomes lax. The protein chains are still sticky, but when there is a lot of heat in the environment the chains move around too much to stick to themselves and each other. The second is to change the amount of salt or acid in the solution. This tends to affect the side chains that have carboxylic acids or amine groups in them, altering how they interact with the surrounding water. Lastly, you can hit the protein with ultraviolet light or other strong radiation. This causes the amino acids themselves to begin to degrade, so that the atoms within them begin rearranging in various chemical reactions.

When any of these things happen, we say that the protein has become **denatured**. A denatured protein loses its ability to perform its biological role. In fact, many serious diseases are a result of protein denaturing or errors in the protein's production that leads to an incorrect structure. We'll talk more about this in later Topic Texts.

Immense Protein Diversity

Clearly, the primary structure of protein chains (the order of the amino acids) determines a great deal about how the overall protein will fold up in water and how the protein is to be used. The more combinations we can manage to put together, therefore, the more versatile an arsenal of protein chemicals that we can make. How many different proteins can we make? Let's see.

For the first amino acid, we could choose any of the twenty amino acids. For the next amino acid, we also could choose from any of the twenty aa's. So, after the second amino acid is added, we have a total of 20 x 20 = 400 different chemical combinations. For the third amino acid, we can again choose from any of the twenty available aa's. That pushes our total combinations to 8000 unique chemicals. Following this pattern, for a protein that is *n* amino acids long, the total number of chemical combinations would be 20^n. So, a simple amino acid chain sixteen aa's long can be made 65536000000000000000 ways. Ultimately, this means that building proteins in such a modular fashion can result in a HUGE potential number of differently shaped chemical blobs, each having their own unique characteristics. That is a LARGE number of shapes and lots of "diners" where chemicals can meet up and interact. This is why proteins serve such a large number of roles in biology.

For the moment, however, we need to pause to consider an important question. Given all the possible amino acids, how do living systems know which amino acids to use and in what order? If we randomly put the amino acids together into a chain, we can certainly arrive at a great number of different structures, but living systems are not random. They seem to "know" what amino acid combinations make useful protein structures and which combinations do not, but how? The answer to this lies in the order of genetic information that exists within the individual. The start of the details begins in the next unit.

Summary

In this section, we discussed proteins.

- Proteins fill many different roles in biological systems: enzymes, scaffolding / cellular structure, contraction and organism movement, cellular transportation, nutrition for young, some hormones and immunity.
- Amino acids are characterized by having an N-C-C backbone, with a unique side group (usually labeled R). The N on the left is an amine group. The C on the right is a carboxyl group. The hydrogen of the carboxyl group is usually snapped off and attached to the amine group.
- There are four broad categories for amino acids: nonpolar, polar, acidic and basic. The nonpolar amino acids are usually tucked inside the protein. The other three types are usually on the outside, facing water.
- Amino acids are attached together using dehydration synthesis into long chains called proteins (or peptides). The order in which they are put together is called the primary structure. Since there are twenty amino acids, there are an astronomical number of possible primary structure combinations that a living system can make.
- Peptide chains are very sticky with lots of polar bonds on its N-C-C backbone. This causes them to take up many strange shapes.
- The secondary structure of a protein is characterized by alpha-helices and beta-sheets.
- The alpha helix is a structure in which the peptide chain spirals into a tube due to the interaction of polar bonds along the N-C-C backbone.
- The beta-sheet is a structure in which the peptide chain doubles over and lines up with itself to form a ruffled chip or wavy roof-like shape.
- The alpha helices, beta-sheets and remaining peptide chain forms into a blob known as the tertiary structure of the protein.
- Multiple peptide chains interacting with each other makes up the quaternary structure of the protein.

- Sulfur-sulfur bonds result when two sulfur containing amino acids (cysteine or methionine) are clipped and connected together via their sulfur atoms.
- Proline bends occur along the peptide chain where proline is. This is due to proline's unusual bent shape that occurs in its N-C-C backbone.
- Discreet areas in a protein are often called domains, and these have their own chemistry they perform. Different proteins can sometimes have very similar domains, because part of their overall function is similar.
- Cellular structural proteins tend to be long and thin, connecting into threads that span long distances.
- Transport proteins are characterized by having two leg-like domains, which "walk" along structural proteins to deliver various molecules to distant parts of the living system.
- Enzymes are shaped so they have nooks in them that only their target molecule(s) fit into. Enzymes speed the reactions of these target molecules.
- Protein structure denatures via heating, salt/acid changes and radiation. Protein denaturation can lead to illness.
- The incredible number of combinations of amino acid chains that living systems can make is stored in the organism's genetics.

Unit 11

Nucleic Acid

> **In this section we cover:**
> 1) The nucleotide building blocks
> 2) DNA structure
> 3) RNA structure
> 4) The Role of DNA as the information molecule
> 5) The Roles of RNA
> 6) The various roles of individual nucleotides

You *cannot* get closer to the heart of biology than the study of nucleic acids. The major role of nucleic acid is to be the genetic material, within which all the information of life is stored. There are other roles that nucleic acids play, however, so we'll need to check those out as well. First, let's take a peek at the monomers of nucleic acids (the **nucleotides**) and how they come together to make the two major forms of nucleic acid: **DNA** and **RNA**.

Nucleotides

Nucleic acids are polymers made up of monomers called nucleotides. The primary structure of nucleotides is fairly complex, so let's first look at a very simplistic design:

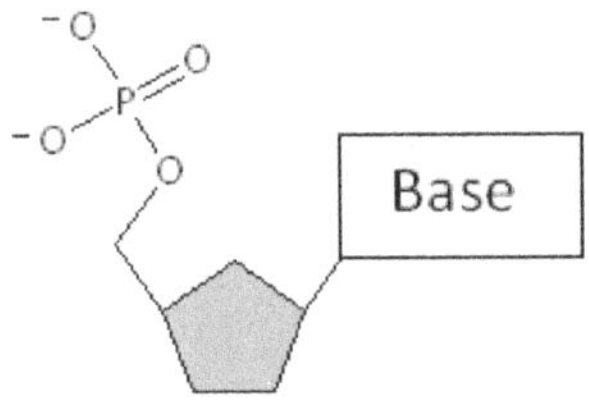

The gray bottom pentagon is a 5-carbon monosaccharide. In the upper left is a phosphate, and in the upper right is a confusing bunch of carbon rings called a "base." Let's slowly unpack the complexity, one part at a time. There are two different types of 5-carbon monosaccharides that can occur in place of the bottom pentagon: the sugar **deoxyribose** and the sugar **ribose**. Deoxyribose has one less OH than ribose. In a nucleotide, the five-carbon sugars would look like this:

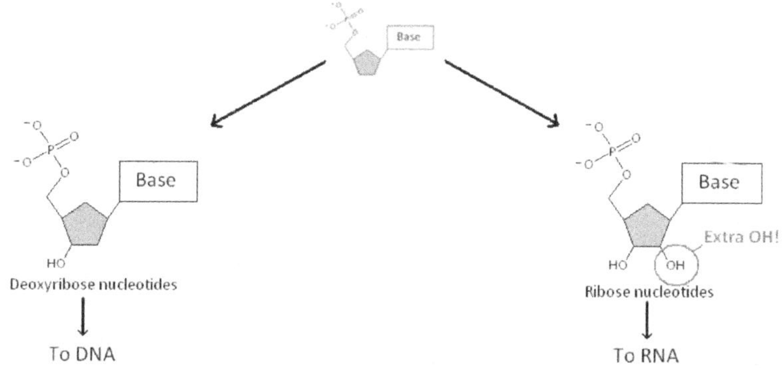

The nucleotides that have deoxyribose (left) are used to make DNA (**deoxyribonucleic acid**). The nucleotides that have ribose (right) are used to make RNA (**ribonucleic acid**). Let's focus first on DNA.

Introduction to DNA Structure

There are four bases that can appear in DNA nucleotides:

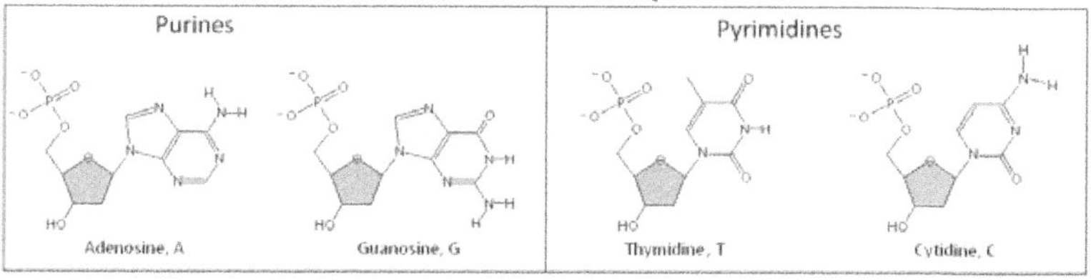

As you can see from this image, the bases are divided into two major groups. Those whose structures contain two rings are called **purines**. The bases with only one ring are called **pyrimidines**. The shorthand notation for these nucleotides is A, G, C and T, for **adenosine**, **guanosine**, **cytidine** and **thymidine**. These nucleotides are strung together to make DNA polymers by dehydration synthesis as shown here on the following page.[15]

[15] Technically, nucleotides start with three linked phosphates (nucleotide triphosphates) and not just one. Two phosphates are removed and one phosphate remains in the nucleic acid backbone. The end result is the same, but the essential pattern is harder to recognize. The important thing to understand is the recurring

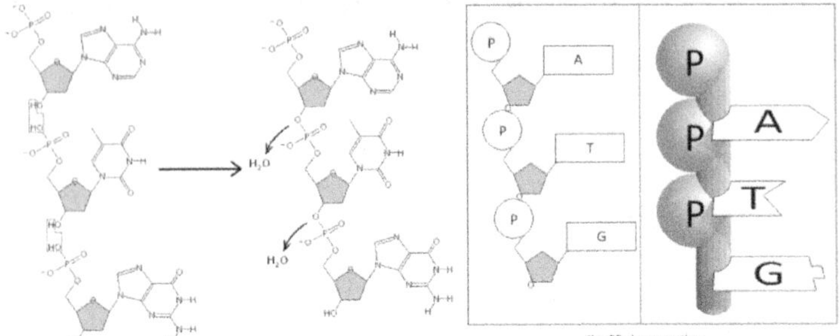

Simplified representions

As you can see from the two rightmost structures, the resulting small polymer looks like half of a ladder, the backbone made from sugar-phosphate repeats. This particular half-ladder DNA chain is only three bases long. The DNA in living systems typically has on the order of billions of nucleotides! DNA bases have an interesting ability to "pair" with each other, and it is the pairing ability of the bases that makes nucleic acids unique, setting the stage for the inheritable nature of biological information. To clarify, let's zoom in on the bases themselves and see how they interact with each other.

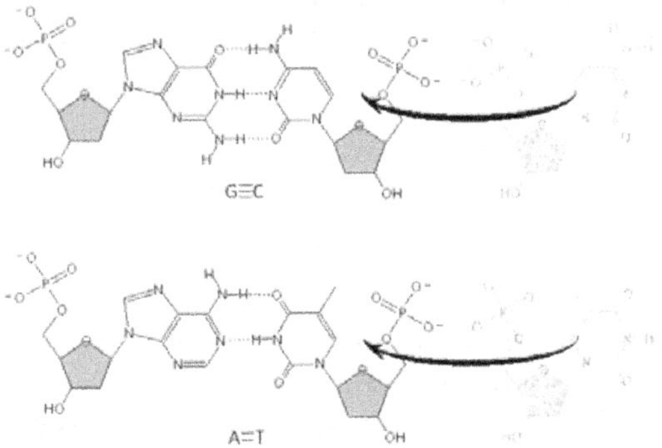

In this last image, I've shown the base for guanosine interacting with cytidine (flipped to face left), and adenosine interacting with thymidine (also flipped to face left). The striking characteristic of these interactions is the hydrogen bonding between the bases (represented by the dotted lines). G has three hydrogen

---

theme of dehydration synthesis. Two monomers are put together and water is released in the process.

bonds that it shares with C, and A has two bonds it shares with T. This bonding pattern is very specific. Because of the specificity of interaction between these two bases, the two are referred to as a **complementary base pair.** You never see a G bonding with a T, or an A bonding with a C, for example[16]. They simply don't fit that way. Due to this pairing nature and to the complexity of nucleotides, I will often draw them in a very cartoonish and stylized way.

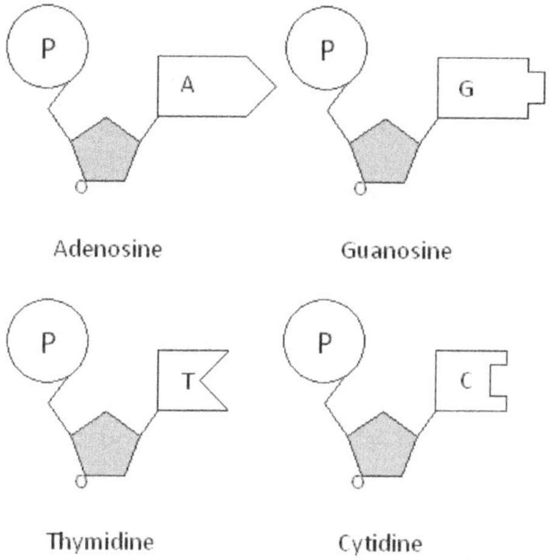

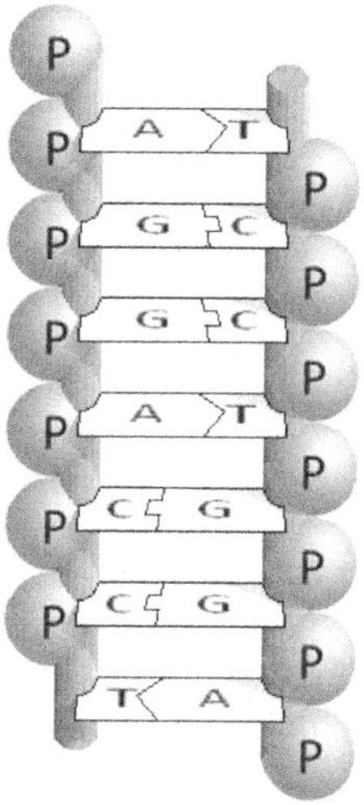

Based on previous images, you might think I've made a mistake with the order of phosphates in the backbones of the nucleotides on the far right. One sugar-phosphate backbone looks like it's going backwards! The sad truth, (and one of the most surprisingly difficult concepts in learning about DNA), is that when two nucleotide chains pair, their backbones end up going in opposite directions. Naturally, it's difficult to predict this outcome based on the simple models of the nucleotides we've looked at, but one must remember that the actual shape of a molecule is often grossly distorted by 2D

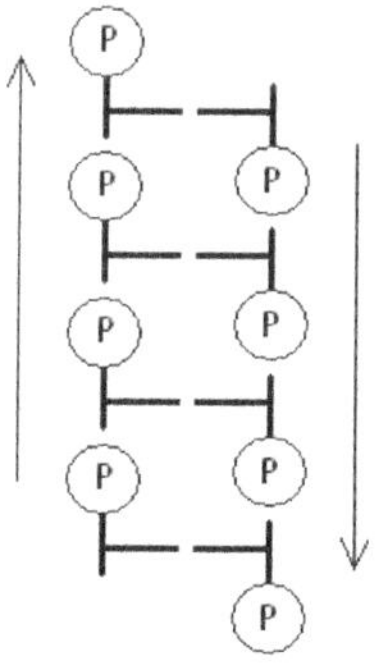

16 ...well, assuming there isn't a major molecular screw-up, that is.

drawings. In the case of nucleotides, the sugar ring, the phosphate and the base are not actually lying flat with each other. When the actual 3-D restraints are taken into account, the only way for the two chains to fit together is for their backbones to go reverse from each other.

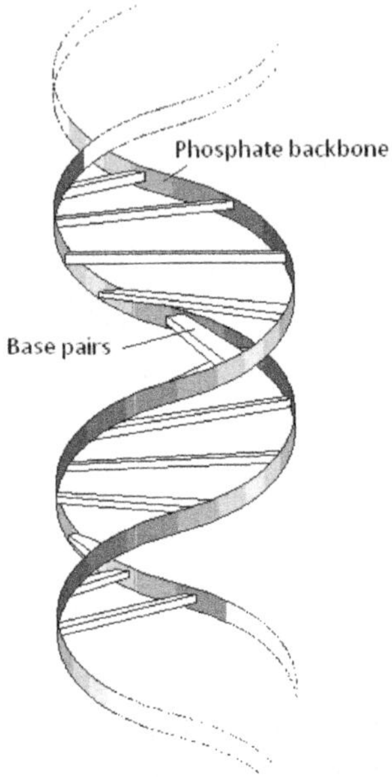

It gets even more complex than this, however, because the paired DNA chains (or DNA **duplex**) twist around each other as well, creating the famous **DNA double-helix**, shown here:

This is how DNA exists inside living systems such as you and me. It is an incredibly long and twisted helical pair of DNA strands that holds all the information for basic biological functions within living systems. In many of the future Topic Texts we will be expanding on the nature and function of this molecule and how its helical structure makes things tricky for living systems. For the moment, however, let's not forget about RNA, because it is equally important in the big picture.

RNA, A First Look

Now that we've had an initial glance at DNA structure, let's take a brief look at RNA. As we mentioned before, the major difference between the RNA nucleotides is the addition of the extra "OH" on the sugar ring at the bottom-right of the nucleotide molecule. Also, the bases in RNA are almost identical to those in DNA, with one exception: the thymidine is replaced by **uridine**. If you look at the following image, however, you might notice that uridine is almost the same as thymidine, except for a missing methyl (CH_3) group. This nearly identical structure to thymidine means that it also pairs with adenosine (i.e., A = T for DNA, and A = U for RNA).

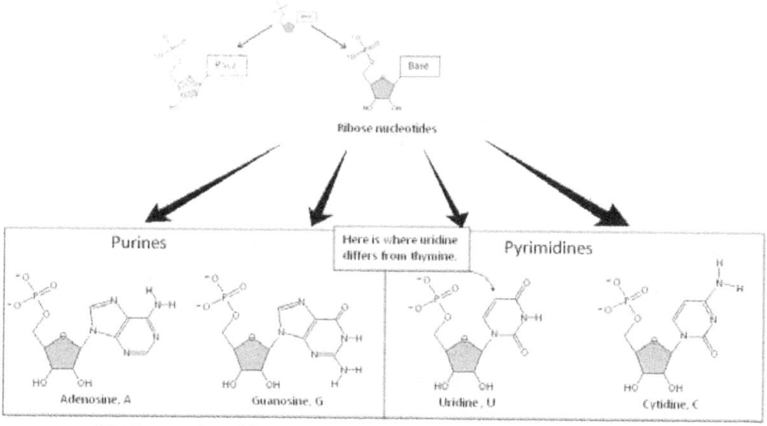

RNA nucleotides are put together into RNA polymers by dehydration synthesis, just like DNA and most other biological polymers. Although RNA can often twist into helices like DNA does, its twisting is not nearly as tight and compact as DNA. The reason is that the extra "OH" tends to get in the way of the helix formation. In addition, RNA strands are not nearly as long as DNA strands, and each one tends to fold up into characteristic formations, with short stretches of helical parts.

As we shall see, the fact that RNA can clump into many weird shapes gives it a wide array of uses within living systems, kind of like what we discussed for proteins in the previous unit.

Now that we've seen the basic structure of nucleic acids, let's transition to get our first glimpse of their many roles in living systems.

The Many Roles of Nucleic Acids

DNA's main role: Genetic Information

Remember the argument about how twenty kinds of amino acids could lead to an incredible number of different protein chains? A similar principle works here as well.

A generic but typical RNA 3-D structure

Let's calculate the number of ways we can put together DNA nucleotides into a chain, starting with the first one. Our first nucleotide can have one of four bases. Therefore, before we begin building the chain, we already start with four total possibilities. We now add the next nucleotide, which can also have any one of the four bases. That gives us 4 x 4 = 16 total possibilities. We add yet another, and once again we can choose among four. This leads us to 4 x 4 x 4 = 64 possible combinations for a three nucleotide chain. As you can see, the pattern of 4^n soon emerges, where n is the number of monomers we've added to the chain. If we have 100 nucleotides, the total number would be 4^{100}, which is well over 1.6×10^{60}. To be perfectly clear how big that is:

100 nucleotides ---> 4^{100} combinations, which equals OVER:

1,600,000,000,000,000,000,000,000,000,000,000,000,000,000,00
0,000,000,000,000,000,000 combinations.

That's a LOT of combinations, and it's just from 100 bases! Still, it's nothing in comparison to the combinations in living systems. The number of DNA nucleotides in the **genome**[17] of any given organism could be in the *billions*. That would give us $4^{billions}$ combinations. I don't think this book is big enough to fit the number that comes out to be.

So what? It's just a bunch of mathemagic, right? Well, the fact that we can have such unimaginable combinations in the arrangement of nucleic acid chains leads to the ability to *store information*. For example, suppose we "read" the chain in segments of three nucleotides at a time. Every three nucleotides can be arranged in 64 ways. So, imagine assigning a letter or punctuation

---

[17] Genome = complete collection of all hereditary genetic information within the nucleus of a cell. You'll learn more about that later.

mark to each combination in the 64 combinations. You could, in theory, send a secret code to a friend by just sending the sequence of nucleic acids.

AGC CAT ATC GGC ACC ATG ATC TAT AGG CGT ACC GTC ATC CTT CAT. . . .

Could be the start of a message like:

"Dear Sam,

I haven't the faintest clue what this idiot is talking about. I need a biology study guide to help me understand this study guide. . ."

Of course, DNA doesn't store messages in English (or any other human language). It stores a code that tells the rest of the living system how to behave on a biochemical level. Specifically, it encodes the order of amino acids in the proteins that keep us alive. DNA is the blueprint for protein combinations! How this code works and is interpreted is the focus of future Topics Texts. For the moment, it is only important to understand the significance of this system: Coded messages, stored as the sequence of base pairs in a long twisted molecule, are copied every time a living system has a child. The child, in time, copies the message as well, passing it along to the next generation, and so forth.

Copying the DNA is very easy, in principle. Since the nucleotides pair up in a very specific way, the living system can pull the DNA ladder slowly apart and build a new strand by matching up the bases on the old strand. In essence, the old DNA is used as a template for the new copy. More on that story much later. . .

Primary roles of RNA: Protein Production

RNA's primary role is in the production of proteins. Basically, RNA carries packets of information from the DNA to other parts of the living system, and these packets tell the living system the order in which to put amino acids together. If you'll recall from the unit on proteins, the order of the amino acids is the primary structure of proteins, and this structure determines the secondary, tertiary and quaternary structure as well. By determining the overall *structure*, the order of amino acids determines the overall *usefulness* of the protein. So, in this sense, RNA acts as a postal worker, carrying the blueprints from the DNA engineer to the factory where the blueprints are used to make the protein product. We'll be covering that topic in another Topics Text, but for a quick look: the order of DNA bases determines the order of RNA bases, and this, in turn, determines the order of amino acids that get put into proteins.

Ribozymes - RNA Enzymes

Proteins are usually the first macromolecule that people think of when the word "enzyme" is mentioned, but some RNA molecules can also serve as enzymes. Whereas DNA is deliberately paired to have complementary strands, (and is thus limited to information storage), RNA is usually made without a complementary strand. They float around in water and fold up into weird shapes, primarily because many of the nucleotides are complementary to others within the *same* chain. As a result, the molecule assembles by folding and twisting, creating specific and elegant structures. It shouldn't be too surprising then that they have little nooks and spaces that can serve as recognition and binding sites for small molecules to fit into. In this manner, some RNA molecules (called "**ribozymes**") act just like protein enzymes, speeding the rates of very specific reactions. Most reactions in living systems are catalyzed by protein enzymes, though; ribozymes are a definite minority as catalysts.

The Other Roles of Individual Nucleotides

Nucleotides have a small number of other roles besides becoming part of large nucleic acid chains. I'll list a few here.

Nucleotides as Energy Carriers

A few nucleotides are involved in energy management. In a way, you can think of a living system as an immensely complex network of chemical reactions that are all trading energy back and forth to each other. One of the most common ways in which reactions exchange energy is in the form of a molecule known as **ATP**, or adenosine triphosphate. ATP is, in a sense, the main energy "currency" of living systems, but it is not the only one used. A few

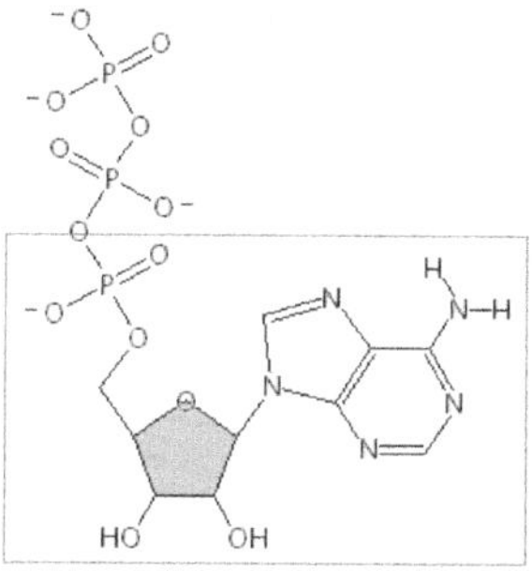

Adenosine triphosphate, ATP

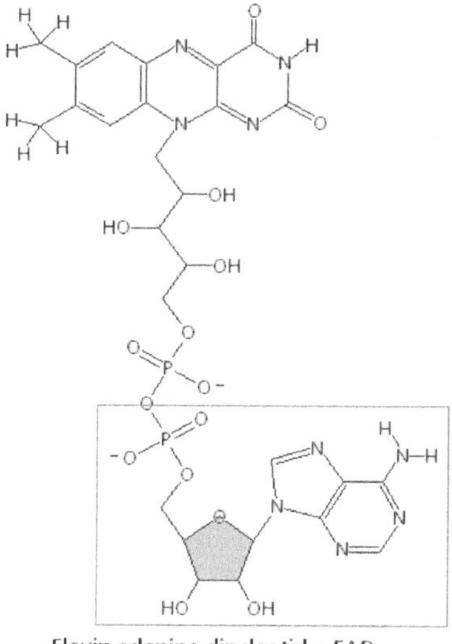

Flavin adenine dincleotide, FAD

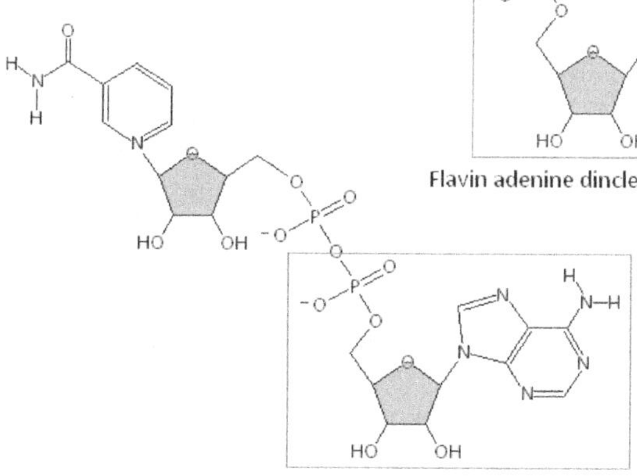

Nicotinamide adenine dinucleotide, NAD^+

other molecules, such as FAD (**flavin adenine dinucleotide**) and NAD^+ (**nicotinamide adenine dinucleotide**) are used as secondary energy currencies, but these are only exchanged between a limited number of reactions.

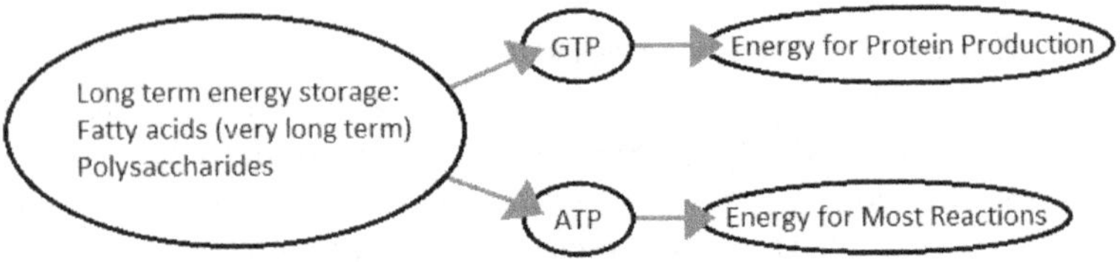

These particular molecules are all derived from adenosine, but the other nucleotides are used as well. **Guanosine triphosphate** (**GTP**), for example, is very similar in structure to ATP, but is used specifically as an energy source for protein production. Another example is **cytidine triphosphate** (**CTP**), which is similar in structure to ATP and GTP. CTP is used primarily as an energy source for both phospholipid production and for attaching small sugar chains onto some proteins (a process called **glycosylation**). We cover that process in a later Topic Text.[18]

Nucleotides as Coenzymes

Some nucleotides are modified to serve as **coenzymes**. Coenzymes (meaning "with enzyme") are extra little molecules that proteins sometimes need to efficiently serve as enzymes. Think of them as assistants to proteins. Not all enzymes need them, but those that do generally cannot work well (or at all) without them. A good example of a coenzyme is Coenzyme A, which is derived from adenosine, shown on the next page.

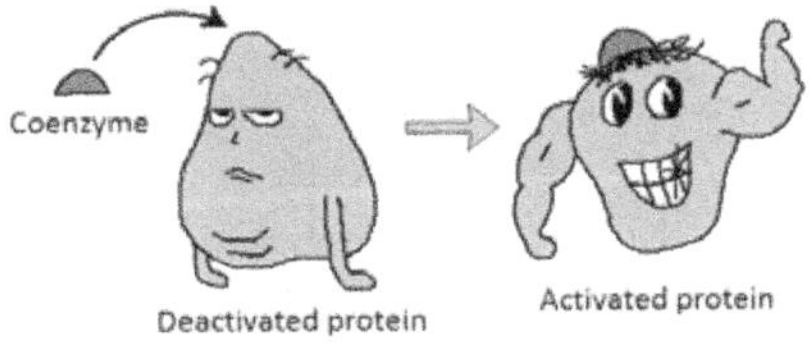

[18] The pattern would suggest the use of TTP, or thymidine triphosphate, as an energy source as well. Oddly enough, TTP is not used in this manner.

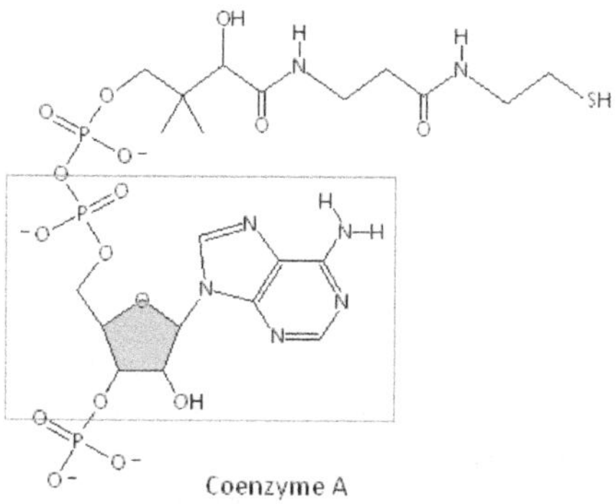

Coenzyme A

Coenzyme A fits into a couple different proteins that speed up the production or breakdown of fatty acids. It's also used in one of the final steps that prepares glucose to be fully broken down for energy in the cell. We'll be learning more about this in later books, specifically during discussions on glycolysis and the citric acid cycle (also called the Krebs Cycle or Tricarboxylic Acid Cycle). A detailed look at fatty acid production and breakdown is usually left for biochemistry courses.

There are lots of different coenzymes found in living systems. Not all are made from nucleotides, and not all nucleotides are converted into coenzymes.

Nucleotides as Intracellular Messengers

A very small number of nucleotides are specifically made to serve as messengers within cells. They act to alert the cell to specific environmental changes, much like a scout might warn an outpost of an impending attack. These molecules usually hang out at the cell surface and start a chain reaction inside the cell when some kind of event happens outside the cell. The classic example of this type of molecule is **cyclic adenosine monophosphate**, or **cAMP**. Another example is **cyclic guanosine monophate**, or **cGMP**.

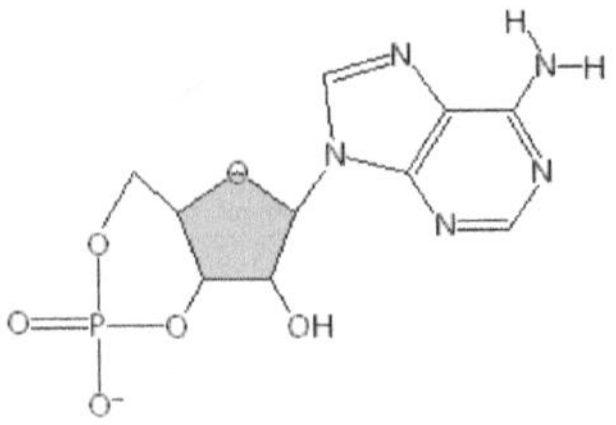

Cyclic AMP

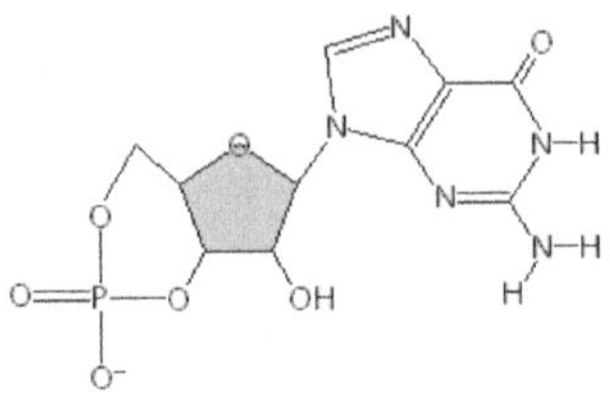

Cyclic GMP

With all of that covered, we've finished our look at the basic chemistry of biology! We are now ready to begin our look at the basic functions of cells. I hope to see you up ahead in the next Topic Text.

Summary

In this section, we looked at the basic structure of nucleic acids, starting from nucleotides and building up to the two main nucleic acid polymers: DNA and RNA. We also discussed the main roles of DNA, RNA and a few of the individual nucleotides.

- Nucleotides are made from a 5-carbon sugar, a phosphate group and a complex carbon-ring structure known as a base.
- Nucleotides are divided into two broad categories depending on the type of 5-carbon sugar they have: deoxyribose nucleotides and ribose nucleotides. Ribose nucleotides differ from deoxyribose nucleotides by the extra –OH group attached to the 5-carbon sugar ring.
- Deoxyribose nucleotides are used to make DNA. They are typically called DNA nucleotides.
- Ribose nucleotides are used to make RNA. They are typically called RNA nucleotides.
- DNA nucleotides can have four possible bases. The molecules that result are: adenosine (A), guanosine (G), thymidine (T) and cytidine (C). Adenosine and guanosine are called purines. Thymidine and cytidine are called pyrimidines.
- Adenosine pairs with thymidine through two hydrogen bonds. Guanosine pairs with cytidine through three hydrogen bonds.
    - DNA:
      A pairs with T   G pairs with C
- RNA nucleotides bases are the same as DNA nucleotide bases, except for thymidine. RNA uses uridine in place of thymidine, but uridine looks just like thymidine except for a missing methyl group. Like thymidine, uridine pairs with adenosine via two hydrogen bonds.
    - RNA:
      A pairs with U   G pairs with C
- Nucleotides in DNA and RNA are connected together via dehydration synthesis to make DNA and RNA chains, respectively.
- Because nucleotides pair with each other, DNA chains form pairs as well. These paired chains twist around each other to make long DNA double helices.

- RNA chains are very short and many tend to curl up into blobs. This behavior is due to the nucleotides pairing to others within the same chain.
- DNA is used primarily for information storage. The sequence of nucleotides in the chain codes for the instructions on how the rest of the living system should behave biochemically.
- RNA has many roles, but its main role is to carry the information stored in DNA to the rest of the biochemical machinery of the living system to make proteins.
- RNA can also serve as enzymes called ribozymes. Like protein enzymes, these work to speed the rate of a very specific number of reactions. This is a very minor role for RNA.
- A few molecules, (most of them derived from adenosine), are used as energy currency within living systems. The most common is ATP.
- A few of the nucleotides, namely adenosine, can be converted into coenzymes, such as coenzyme A. A coenzyme is a small molecule that helps enzymes function properly.
- A few molecules derived from various nucleotides are used as messengers, informing cells of changes in chemical circumstances outside the cell. cAMP and cGMP are the most important examples.

A

B

C

D

E

F

G

H

I

K

L

M

N

O

P

Q

R

S

T

U

V

W